D0919634

HANDBOOK OF HIGH RESOLUTION MULTINUCLEAR NMR

Handbook of High Resolution Multinuclear NMR

C. BREVARD
Bruker Spectrospin
Wissembourg, France

P. GRANGER
Université de Rouen
Mont St Aignan, France

A Wiley-Interscience Publication
John Wiley & Sons New York Chichester Brisbane Toronto

Library of Congress Cataloging in Publication Data:

Brevard, C.
Handbook of high resolution multinuclear NMR.
"A Wiley-Interscience publication."
Bibliography
Includes index.
1, Nuclear magnetic resonance spectroscopy.
I. Granger, P. (Pierre) II. Title.

QD96.N8B75 543′.0877 81-8603
ISBN 0-471-06323-1 AACR2

Printed in the United States of America

10 9 8 7 6 5 4 3 2 1

We do not remember days,
only moments

PREFACE

The widespread availability of dedicated multinuclear high-resolution NMR spectrometers has opened up a new and exciting field in chemical research for the organic and inorganic chemist. A great many problems can now be tackled in the NMR study of "exotic" nuclei such as tungsten, oxygen, silicon, lead, silver, and cadmium. Actually, the entire periodic table has become accessible since the development of high or very high field, wide-bore multinuclear FT spectrometers.

It has been our goal in writing this book to implement initial NMR experiments on nuclei other than proton, fluorine, phosphorus or carbon 13. To begin with the experimenter will require any number of hints, among which might be the exact resonance frequency of a readily available compound or a clearcut chemical shift scale for the isotope already under study.

Of course, relaxation times, NOE effects, acquisition parameters (and folded lines!) are important in such experiments because most of these low sensitivity nuclei can be observed only in the Fourier mode: we describe steps fully in the first part of the book; the second part is a nucleus—one page on which to build.

It is our hope that this monograph will obviate too much time from being spent in cumbersome spectrometer tuning and will prepare the chemist to start at once with the NMR study of choice.

C. Brevard
P. Granger

Liebfrauenberg, France
March 1981

ACKNOWLEDGMENTS

We should like to record our indebtedness to all people who provided us with unpublished results or rare samples, among them, we mention

A. Bardy, J. P. Beaucourt, W. E. Hull, J. P. Kintzinger, Le Laboratoire de Chimie Minérale de l'Université de Rouen; B. Lindman, R. Marchand, J. Moulines, Mme M. Postel, W. von Philipsborn, L. Pichat, and Mme M.C. Vitorge-Pagnier. P. Granger greatly acknowledges the Etablissement Public Régional de Haute Normandie for financial support.

Miss M.C. Hermann typed the entire manuscript with endless patience.

C.B.
P.G.

CONTENTS

USUAL CONSTANTS (MKSA)

From Handbook of Chemistry and Physics, 60th Edition, 1979-1980.

Avogadro's constant,	$N = 6.022045\ 10^{23}$ mole^{-1}
Planck's constant,	$h = 6.626176\ 10^{-34}$ J.s.
$\hbar$	$h/2\pi = 1.054589\ 10^{-34}$ J.s.
Boltzman's constant,	$k = 1.380662\ 10^{-23}$ J.K.$^{-1}$
Magnetic permeability of vacuum,	$\mu_o/4\pi = 10^{-7}$ m.kg.C^{-2}
Elementary charge,	$e = 1.6021892\ 10^{-19}$ C
Electron rest mass,	$m_e = 9.109534\ 10^{-31}$ kg
Proton rest mass,	$m_H = 1.6726485\ 10^{-27}$ kg
Bohr magneton,	$\mu_B = 9.274078\ 10^{-24}$ J.T.$^{-1}$
Nuclear magneton,	$\mu_N = 5.050824\ 10^{-27}$ J.T-^{1}
Proton moment,	$\mu_H = 1.4106171\ 10^{-26}$ J.T.$^{-1}$
Gyromagnetic ratio of proton,	$\gamma_H = 2.6751987\ 10^{8}$ rad $s^{-1}T^{-1}$
$\gamma_H/2\pi$,	$= 4.258222\ 10^{7}\ s^{-1}T^{-1}$
Electric permittivity of vacuum :	$\varepsilon_o = (1/36\pi)\ 10^{-9}m^{-3}kg^{-1}s^{2}C^{2}$

SYMBOLS AND ABBREVIATIONS

A	Hyperfine coupling constant
a%	Natural abundance
ADC	Analog to digital converter
AQT	Acquisition time
$\vec{B}_o$	Main induction
$\vec{B}_1$	Observing radiofrequency field
$\vec{B}_2$	Decoupling radiofrequency field
$C_{//}, C_{\perp}$,	Components ot the spin rotation tensor
D_{ij}	Direct coupling constant
e	Electron charge
FID	Free induction decay
FT	Fourier transformed
$\Delta G^{\neq}$	Free energy of activation
$\hbar$	$h/2\pi$ = reduced Planck constant
I	Eigenvalue of the angular spin momentum

I_x, I_y, I_z	Eigenvalues of the component of I
$I_{//}$, $I_{\perp}$	Components of inertial moment
$^nJ_{A-X}$	Indirect spin-spin coupling constant
$^nK_{A-X}$	Reduced indirect coupling constant
M, Mo, M_∞	Total magnetic moment
M_x, M_y, M_z	Components of M.
M_{eff}	Total magnetization along $\vec{B}_o + \vec{B}_1$
N, n	Number of scans
NOE	Nuclear Overhauser Enhancement
Q	Quadrupole moment
q	Electric field gradient
R_A	Receptivity of nucleus A
RF	Radiofrequency field
RFS	Repetitive frequency shift
r_{ij}	Distance between spin i and j
S	Eigenvalue of the spin momentum of the electron
$\Delta S^{\neq}$	Entropy of activation
SW	Sweep width
S/N	Signal to noise ratio
T	Temperature in°K
T_1	Spin lattice relaxation time
$T_{1\rho}$	Relaxation time in the rotating frame
T_2	Spin-Spin relaxation time
T_2^*	Apparent spin-spin relaxation time

t	Time interval
α or β	Pulse angle
γ_A	Gyromagnetic ratio of nucleus A
δ	Chemical shift in ppm
η	NOE enhancement factor
η_a	Asymmetry parameter of an electric field tensor
θ	Pulse angle
μ	Electronic magnetic moment
$\mu_o/4\pi$	Magnetic permeability of vacuum
ν	Resonance frequency
ν_{ref}	Resonance frequency of a reference line
ν_o	Frequency of the applied RF field
$\Delta\nu_{1/2}$	Half-line width
σ	Screening constant
$\sigma_{//}$, $\sigma_{\perp}$	Components of the screening tensor
τ	Time interval
τ_c	Correlation time
χ or χ_v	Volumic susceptibility
ω	Angular velocity

HANDBOOK OF HIGH RESOLUTION MULTINUCLEAR NMR

PART ONE

CHAPTER 1

NMR PARAMETERS

When Fourier transform NMR experiments are performed on nuclei other than proton, fluorine, or carbon 13 every experimentalist is faced with a "recording strategy" that requires a knowledge and clear definition of NMR parameters such as sensitivity, natural abundance, NOE effects and T_1 or T_2 relaxation mechanisms. These parameters are defined[1] and exemplified as broadly as possible in this chapter.

1.1 SENSITIVITY, NATURAL ABUNDANCE, AND RECEPTIVITY

1.1.1 Sensitivity

At constant induction B_o, the resonance strength of a given isotope X, is proportional to

$$K \underbrace{\gamma_X^3 \, I_{(x)} \, (I_{(x)} + 1)}_{\text{sensitivity}}$$

Where $I_{(x)}$ = isotope spin quantum number,

γ_X = isotope nuclear magnetogyric ratio.

1.1.2 Natural Abundance

Natural abundance is defined as

$$a_X\% = \frac{\text{number of nuclei (given isotope) in the sample}}{\text{total number of nuclei (natural element) in the sample}}$$

1.1.3 Receptivity

$$R_X = a_X \cdot \gamma_X^3 \cdot I_{(X)} \cdot (I_{(X)} + 1)$$

This is the parameter to be considered in any observation of X isotopes. If enriched samples are used, a_X is replaced by the isotope enrichment value.

1.1.4 Relative Receptivity

$$R_X^Y = \frac{a_X \gamma_X^3 I_{(X)} (I_{(X)} + 1)}{a_Y \gamma_Y^3 I_{(Y)} (I_{(Y)} + 1)}$$

This formula compares X isotope receptivity with a standard Y nucleus. As an example,

$$R_{^{195}Pt}^{^{13}C} = 19.5$$

means that a ^{195}Pt signal should be 19.5 stronger than a ^{13}C signal for the same molarity. Of course, this relative receptivity does not take into account the dynamic effects or different relaxation mechanisms that can drastically modify signal-over-noise ratios under identical recording conditions for X and Y species.

1.2 CHEMICAL SHIFTS

1.2.1 Definition

$$\delta_{ppm} = \frac{\nu - \nu_{ref}}{\nu_{ref}} \cdot 10^6 = \frac{\Delta\nu \text{ (Hz)}}{\nu_{ref} \text{ (MHz)}}$$

Where ν = resonance frequency of the observed line,
ν_{ref} = resonance frequency of a reference line,
$\Delta\nu$ = $(\nu - \nu_{ref})$,

More information on chemical shift will be found in Refs. B10, Vol.2, and B31, Vol.2.

1.2.2 Chemical Shift Sign

The IUPAC convention clearly defines chemical shifts scales

(1,2). The following recommendations are made :

$\delta > o$ for any line detected at a higher frequency than the chosen reference (deshielding, low field);
$\delta < o$ for any line detected at a lower frequency than the chosen reference (shielding, higher field).

Because all Fourier spectrometers are similarly constructed, experimental results are always presented as

$$\overleftarrow{\nu} \qquad \nu_{ref}$$

ppm $\quad \delta > o \quad o \quad \delta < o$

Remark. For heteronuclei the referencing resonance is generally concentration - and temperature-dependent. These two parameters must be clearly defined.

1.3 COUPLING CONSTANTS

1.3.1 Direct Coupling Constant (B8, Vol.1, p.1; Vol.9, p.1; B10, Vol.2)

$$D_{ij} = \frac{-1}{4\pi} \hbar\gamma_i\gamma_j < \frac{3\cos^2\theta_{ij}-1}{r_{ij}^3} > \text{(in Hz)}$$

Where θ_{ij} is the angle between the induction $\vec{B}_o$ and the interconnecting vector (modulo r_{ij}) that links nuclei i and j; $<\ >$ indicates a value averaged over all nonrandomly oriented magnetic (i,j) dipoles. Therefore, D_{ij}

(a) is nulled in solution ($<3\cos^2\theta_{ij}-1> = 0$);
(b) is measurable only in anisotropic systems such as solids, nematics, and smectics;
(c) is the subject of molecular geometry studies (B8, Vol.1, p.1) and relates to structural informations gathered on mesomorphic phases (B8, Vol.9, p.1)

1.3.2 Indirect or Scalar Coupling Constants (B21, Vol.7, p.246; B31, Vol.1, p.149)

Consider two nuclei A and B,

$${}^{n}J_{AB} = k\gamma_A\gamma_B \quad \text{(in Hz)}$$

where n = number of bonds between A and B,
k = depends on the A-B bond electronic structure and is often parameterized by the angular geometry of the A-B bond (Karplus type curves).

It should be noticed that $^nJ_{AB}$ does not depend on B_o and may be positive or negative, according to the signs of k, γ_A, γ_B. Sign determination is discussed in Section 3.8.1,2.

1.3.3 Reduced Coupling Constants

In a comparison of the relative magnitude of $^nJ_{XA}$ and $^nJ_{YA}$ a reduced coupling constant $^nK_{MA}$ is introduced (M = X or Y) :

$$^nJ_{MA} = h\left(\frac{\gamma_M}{2\pi}\right)\left(\frac{\gamma_A}{2\pi}\right)\ ^nK_{MA}$$

where $^nK_{MA}$ is expressed in $NA^{-2}m^{-3}$ (newtons.ampere^{-2}.m^{-3})

1.4. DIRECT MEASUREMENTS OF $^nJ_{AB}$

Frequently the $^nJ_{AB}$ indirect couplings must be measured from recorded spectra. Care must be used in measurements of large couplings such as $^1J_{^{117}Sn-^{195}Pt}$ which can amount to 30 kHz! or in nontrivial situations derived from natural abundance isotopic distribution in the compounds under study.

1.4.1 Strongly Coupled Systems of High-Order Spectra

When δ_{AB} (in Hz) is of the same order or smaller than $^nJ_{AB}$ computer analysis of the spectrum is generally required (B2; B7; B8, Vols.5 and 6; B9, Vol.1, p. 280; B10, Vols.1,3 and 10; B21, Vol.4, 104).

1.4.2 First-Order Spectra

δ_{AB} (in Hz) >> any $^nJ_{AB}$.

<u>Observed A Nucleus Coupled with High Natural Abundance (> 90 %) B Nuclei.</u>

If no severe line broadening caused by relaxation or exchange effects, is observed, the coupling pattern of an A nucleus with m.B. equivalent nuclei is a multiplet; the number of lines is given by $2m.I_B+1$ (Figure 1). The relative intensities of the multiplet components are tabulated for P equivalent-spin 1/2 B nuclei.

P	Relative intensities
1	1 1
2	1 2 1
3	1 3 3 1
4	1 4 6 4 1
m	$1 \quad \frac{m!}{(m-1)!} \quad \frac{m!}{2!(m-2)!} \cdots \frac{m!}{k!(m-k)!} \cdots 1$

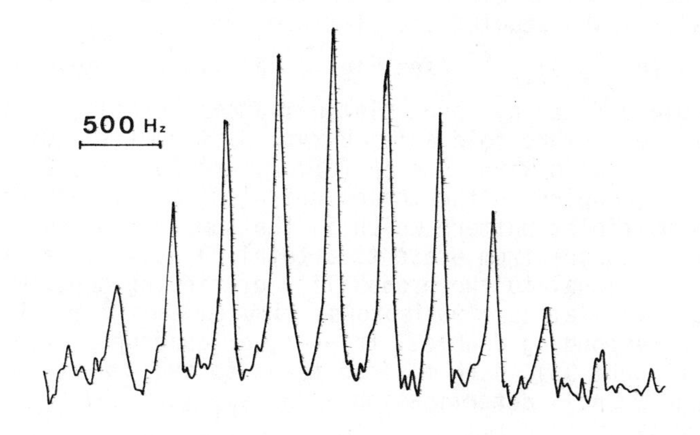

Figure 1. ^{129}Xe spectrum of $(XeF_6)_4$. One xenon nucleus is coupled to 24 equivalent fluorine atoms. Not all the theoretically allowed transitions are detected (4).

For a spin 1/2 isotope coupled to a quadrupolar nucleus ($I \geqslant 1$) relative intensities and line shapes of the spin 1/2 multiplet components have been analyzed theoretically by

Kubo and Suzuki (3). In heteronuclei NMR useful information can be obtained from $^{n}J_{AB}$, especially with spin 1/2 isotopes or for quadrupolar nuclei with low quadrupole moment. It is also worth noting that if the B atoms to which the observed A nucleus is coupled are nonequivalent and give rise to a second-order spectrum, a complete computation of the coupling pattern or a higher field spectrum is necessary to extract long-range couplings. Direct measurement, for example, of $^{n}J_{^{13}C-H}$ in the $Ha{-}C_1 - C_2{-}Hb$ fragment yields $^{1}J_{C_1H_a}$ = 131.44 Hz and $^{2}J_{C_1H_b}$ = 2.6 Hz when $\nu_o\delta_{HaHb}$ = 2 Hz; in fact, real ^{1}J and ^{2}J values amount to 130 Hz and 4 Hz respectively, which corresponds to a 35 % error in the directly measured ^{2}J! A more dramatic situation is illustrated in Figure 2.

Observed A Nucleus Coupled to nB spin 1/2 Nuclei of Intermediate Natural Abundance (20 % < a < 80 %.) In this case the spectrum will appear as the sum of several subspectra whose relative intensities are a function of the natural abundance of the B nuclei. A nice example is given by C. Brown et al.(5) who studied the platinum 195 (a = 33.7 %) spectrum of $(Pt_9(CO)_{18})^{2-}$ (see Figure 3a). In this molecule each triangle defined by three platinum atoms rotates freely around the pseudo-threefold axis; B-type platinum is equivalent; A-type platinum also with $\delta B \neq \delta A$ (cf. Figure 3b). The B-type platinum coupled to the three equivalent A-type neighbors presents a multiplet pattern which is the sum of four subspectra. Each subspectrum presents a total relative intensity that is proportional to the probability of finding one, two, or three active platinum 195 isotopes simultaneously on sites A with a corresponding doublet, triplet, or quadruplet fine structure (Figure 3c). Care must be taken to avoid a straightforward (and wrong!) determination of $J_{^{195}Pt-^{195}Pt}$ by measuring the interval between two lines of the B multiplet in the experimental spectrum.

Observed A Nucleus Coupled to Low Abundance nB Spin 1/2 Nuclei (a^{B} < 20 %). The observed A spectrum appears as one intense line flanked by a doublet (relative intensity n.a^{B} %) whose interval is exactly $^{n}J_{A-B}$. As an example, the ^{13}C proton decoupled spectrum of the methylene moiety of

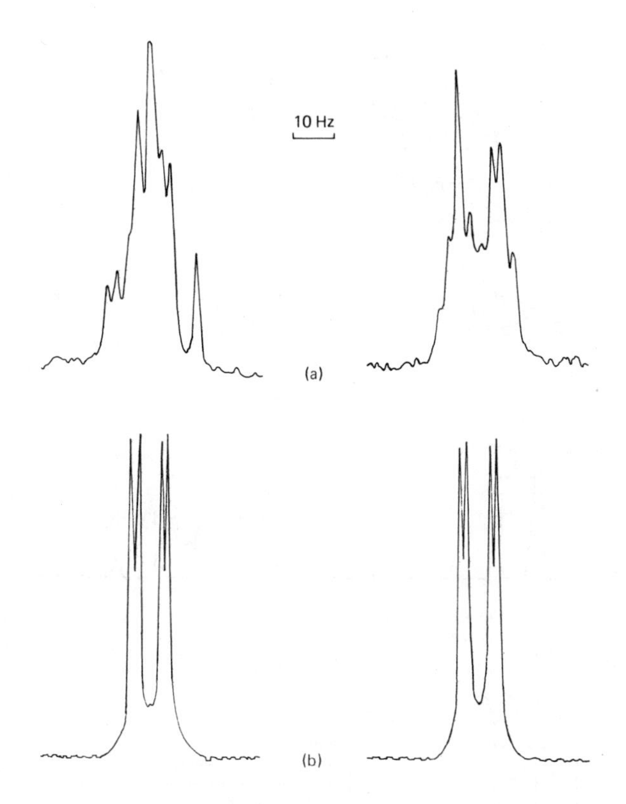

Figure 2. $^{13}C-^{1}H$ coupled spectrum of 2 pyridone (C-3 carbon) recorded at 20 MHz (a) and 75 MHz (b).(From M.C. Vitorge-Panier, These 3ème cycle. Paris 1980.)

$(C_6H_5-Se)_2-CH_2$ shows a single line and a symmetrically disposed doublet. Because of the low natural abundance of selenium 77(a = 7.5 %), the triplet structure due to the low number of $C_6H_5{}^{77}Se-{}^{13}CH_2-{}^{77}SeC_6H_5$ molecules in the solution, although present, is under spectrometer detection capability for a reasonable recording time.

1.5. RELAXATION TIMES: NUCLEAR OVERHAUSER ENHANCEMENT (NOE)

1.5.1 Longitudinal or Spin Lattice Relaxation Time T_1

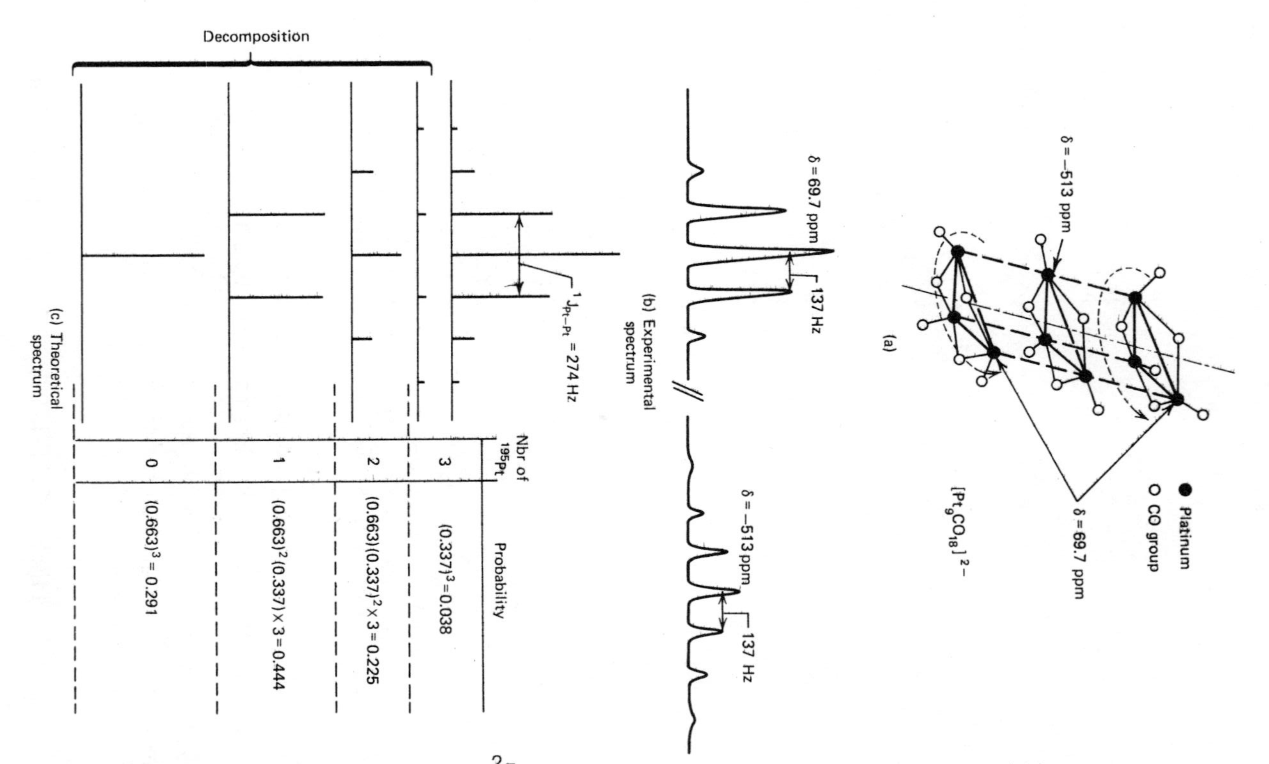

Figure 3. ^{195}Pt spectrum of $(Pt_9CO_{18})^{2-}$ and its decomposition according to ^{195}Pt natural abundance (33.7%) (5). a) the molecular structure, b) the experimental spectrum, c) the theoretical spectrum obtained as the sum of the different subspectra.

Time constant T_1 characterizes the rate of recovery of the z component (M_z) of the total magnetization toward its equilibrium value (M_o) along B_o. This is an energy effect that can also be viewed as the rate of energy transfer from the spin system to the lattice. For an isolated spin

$$\frac{dM_z(t)}{dt} = -\frac{M_z(t)-M_o}{T_1}$$

The experimental measurement of T_1 is developed in Chapter 4.

1.5.2 Transverse or Spin-Spin Relaxation Time T_2

This time constant characterizes the rate of loss of phase coherence of a given spin system, after excitation, in a plane perpendicular to B_o, which leads to a decrease in the y(x) component $M_y(M_x)$ of the total magnetization. This is an entropy effect. For an isolated spin

$$\frac{dM_y(t)}{dt} = -\frac{M_y(t)}{T_2} \quad \text{(same for the x component)}$$

T_2 is directly related to the resonance line width. For a Lorentzian line

$$\Delta\nu_{1/2} = \frac{1}{\pi T_2}$$

where $\Delta\nu_{1/2}$ measures the line width at half-height (in Hz).

Measuring T_2 requires exact experimental conditions (cf. Section 4.1). There is always $T_2 \leqslant T_1$.

1.5.3 Relaxation Time in The Rotating Frame $T_{1\rho}$

When locked after excitation in a strong radiofrequency field B_{eff}, the spin system total magnetization M_{eff} decreases:

$$\frac{dM_{eff}(t)}{dt} = \frac{1}{T_{1\rho}} \left(M_{eff}(0)-M_{eff}(t)\right)$$

When the resonance condition is met, that is, $B_0 - \omega/\gamma = 0$, $T_{1\rho} = T_2$. Generation of B_{eff} does require an intense RF field that imposes special probe design and limited irradiation periods; $T_{1\rho}$ measurements are dealt with in Section 4.6.

1.5.4 The Nuclear Overhauser Effect (NOE) (see B23)

Definition Two nuclei, A and X_i, when close enough, can interact through space in a dipole-dipole mechanism. The irradiation of either one of these X nuclei can lead to a modification of the energy level distribution of the second (A) which results in a modification of the intensity of the resonance line of the A nucleus. This phenomenon, known as the nuclear Overhauser effect (NOE), depends on the observing field and the mobility in solution of the molecule under study (Figure 4). Its measurement is discussed in Section 4.1.

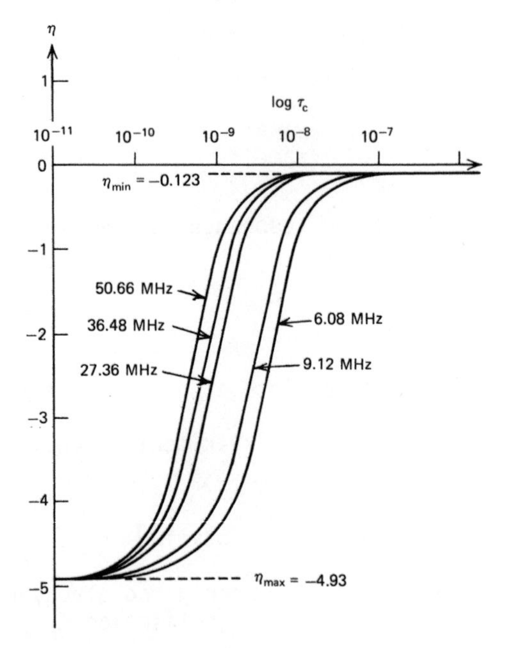

Figure 4. Variation of the NOE enhancement factor η as a function of the correlation time τ_c in ^{15}N-{H} experiments for different observing frequencies.

The NOE is characterized by an enhancement factor

$$\eta = \frac{I - I_o}{I_o}$$

where I_o = a line integral without perturbation of nucleus A,
I = a line integral of nucleus A when irradiating neighboring X nuclei.

In the extreme narrowing condition ($\nu_A \cdot \tau_{cA} << 1$) generally met in high resolution experiments

$$\eta_{max} = \frac{1}{2} \frac{\gamma_X}{\gamma_A}$$

γ_X: irradiated nucleus

γ_A: observed nucleus

Thus η_{max} may be homonuclear, with

$$0 \leqslant \eta \leqslant 1/2$$

or heteronuclear with

$\eta \geqslant 0$ if γ_X, γ_A are of the same sign

$\eta \leqslant 0$ if γ_X, γ_A have opposite signs (see Figure 5)

where $\eta = -1$ nulls the A resonance. This rather troublesome situation may be overcome by using a special pulse sequence (Section 4.1; B20, p. 231) or quenching the A-X dipole-dipole interaction with an inert paramagnetic reagent such as Cr^{III} (acac)3.

<u>Use and Interest</u> If $\eta > 0$ or $\eta < -2$, one clearcut application of NOE speeds up the measurements, because any increase in signal over noise will cause a corresponding drop in accumulation time which amounts to the square of the signal-over-noise enhancement. This relationship should be remembered each time an experiment is carried out on a low γ, low sensitivity spin 1/2. Even if this nucleus is not directly linked to protons, it could help to develop a potential intermolecular NOE enhancement by dissolving the compound in a protonated solvent and accumulating under proton broad-band decoupling conditions (Figure 6).

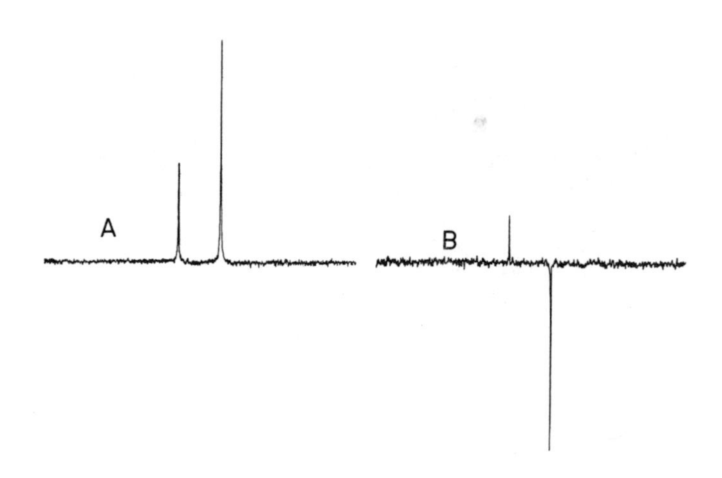

Figure 5. X-{H} intramolecular heteronuclear NOE: > 0 (A) ^{13}C observation, sample C_6H_6. η_{max}^{13C}: 1.989; < 0 (B) ^{15}N observation, sample formamide. η_{max}^{15N}: - 4.933

Because of its dependence on the closeness of A and X spins, NOE can be used for conformationnal (B23) or dynamic (B13; B23) studies after a careful removal of dissolved oxygen in the solution.
This allows the determination of the dipole-dipole relaxation contribution to the different relaxations mechanisms by using

$$T_{1DD} = \frac{\eta_{max}}{\eta_{obs}} T_{1obs}$$

1.6. THE RELAXATION MECHANISMS

Any measured $T_1(T_2)$ relaxation time for nucleus A is related to several potential relaxation pathways.

Hence

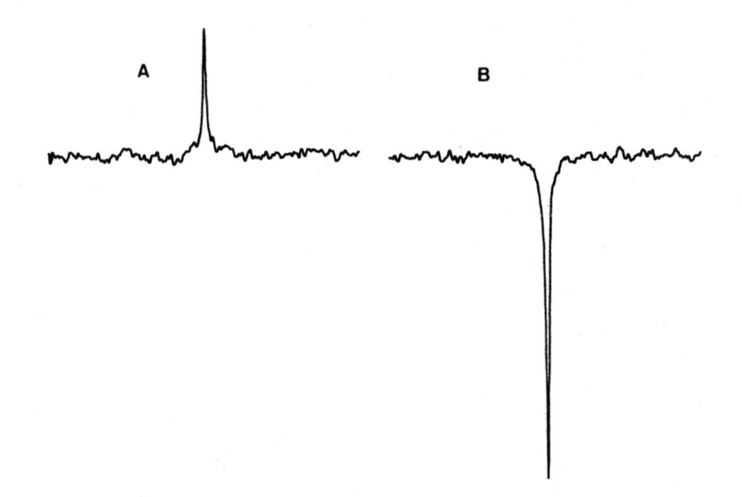

Figure 6. X-{H} intermolecular heteronuclear NOE: sample $Y(NO_3)_3$, M in H_2O. (A) without and (B) with H_2O protons irradiation; $\eta_{max}^{^{89}Y} = -10.2$, $\eta_{exp}^{^{89}Y} = -3.6$.

$$\frac{1}{T_{imeas.}} = \frac{1}{T_{idd}} + \frac{1}{T_{ie}} + \frac{1}{T_{icsa}} + \frac{1}{T_{isc}} + \frac{1}{T_{isr}} + \frac{1}{T_{iQ}}$$

where i stands for 1 or 2 and the various subscripts define a precise relaxation process developed in the following sections (all formulas in SI units).

1.6.1 Dipole-Dipole (dd)

$$\frac{1}{T_{idd}} = \left(\frac{\mu_o}{4\pi}\right)^2 \gamma_A^2 \gamma_X^2 \hbar^2 I_X(I_X+1) \frac{1}{r_{A-X}^6} f_i(\nu_A, \nu_X, \tau_c^A)$$

where

r_{A-X} is the distance between the A nucleus and any surrounding X nucleus (in meters);

$f_i(\nu_A, \nu_X, \tau_c^A)$ is a complex function of correlation time

τ_c^A characteristic of the random motion of the A-X vector and of the ν_A, ν_X resonance frequencies; in the extreme narrowing case $T_{1dd} = T_{2dd}$ and $f(\nu_A, \nu_X, \tau_c^A)$ is simply equal to

$2\tau_c^A$ if A and X are identical and belong to the same molecule,

$4/3\tau_c^A$ if A and X are different but belong to the same molecule,

$4/15\pi r_{A-X}^6 N_A/(p_A \cdot D_A)$ if A and X are identical but belong to different molecules,

$8/45\pi r_{A-X}^6 (N_X/p_X D_X)$ if A and X vary and belong to different molecules,

The first two situations give rise to an intramolecular dipole-dipole relaxation mechanism; the last two develop an intermolecular dipole-dipole relaxation mechanism in which $N_{A(X)}$ represents the number of A(X) spins per m^3, $p_{A(X)}$ is the radius of the molecule (assumed to be spherical), and $D_{A(X)}$ measures the diffusion coefficient in m^2/s. A further approximation can be introduced by the solution viscosity v_s (in $N.s/m^2$). Then

$$\frac{4\pi N_A}{15 p_A D_A} = \frac{8}{5}\pi^2 \frac{N_A v_s}{kT}$$

and

$$\frac{8\pi N_X}{45 p_X D_X} = \frac{16}{15}\pi^2 \frac{N_X v_s}{kT}$$

It should be remembered that τ_c decreases with temperature and therefore T_{1dd} increases.

1.6.2 Electron-Nuclear Relaxation Mechanism (e)

This relaxation mechanism occurs when paramagnetic species (or impurities!) are present in the solution. It is an effecient mechanism that generally overtakes all the others.

It can operate by:

A dipolar interaction,

$$\frac{1}{T_{1ed}} = \left(\frac{\mu_o}{4\pi}\right)^2 \frac{\gamma_A^2 < \mu^2 >}{r^6} f_{ed}(\nu_A,\nu_e,\tau_c^A)$$

where $<\mu^2>$ represents the mean value of the square of the electronic magnetic moment for the paramagnetic species; for example,

$$\mu = 3.11.10^{-23}\ J\ T^{-1} \text{ for } Ni^{2+}$$

$$\mu = 4.64.10^{-23}\ J\ T^{-1} \text{ for } Co^{2+}$$

Generally f_{ed} (ν_A, ν_e, τ_c^A) is different for T_{1e} and T_{2e} (6), but in the extreme narrowing case $T_{1e} = T_{2e}$ and f_{ed} $(\nu_A, \nu_e, \tau_c^A) = 4/3\ \tau_c^A$ with $1/\tau_c^A = 1/\tau_r^A + 1/T_{1e}$, where

$$\tau_r^A = \frac{4\pi v_s P_A^3}{3kT}$$

and T_{1e} is the spin lattice relaxation time of the electron. This formula may sometimes be nonvalid with high field spectrometers.

A contact interaction. This term is obtained from scalar coupling between the A nucleus and the electron:

$$\frac{1}{T_{iec}} = 4\pi^2 A^2\ S(S + 1)\ f_{iec}(\nu_A, \nu_e, \tau_c^A)$$

where A = the hyperfine coupling coupling constant in Hertz,

S = the total electron spin of the paramagnetic center (e.g., S = 1 for Ni^{2+}, S = 3/2 for Co^{2+}

In the extreme narrowing case

$$f_{1ec}\ (\nu_A, \nu_e, \tau_c^A) = \frac{2\tau_{2e}}{3}$$

$$f_{2ec}(\nu_A, \nu_e, \tau_c^A) = \frac{1}{3}(\tau_{1e} + \tau_{2e})$$

where τ_{1e} and τ_{2e} are the correlation times that correspond to the electronic T_1 and T_2, ($\tau_{1e} = T_{1e}$, $\tau_{2e} = T_{2e}$).

1.6.3 Chemical Shift Anisotropy Mechanism (csa)

This mechanism develops from a fluctuation of a nonsymmetrical screening tensor in the magnetic field. The correlation time τ_{ca} describes this random fluctuation. In the extreme narrowing case and for axial symmetry

$$\left.\begin{aligned} \frac{1}{T_{1ca}} &= \frac{40}{3}\left(\frac{\mu_o}{4\pi}\right)^2 \gamma_A^2 B_o^2 (\sigma_{//} - \sigma_{\perp})^2 \tau_{ca}^A \\ \frac{1}{T_{2ca}} &= \frac{140}{9}\left(\frac{\mu_o}{4\pi}\right)^2 \gamma_A^2 B_o^2 (\sigma_{//} - \sigma_{\perp})^2 \tau_{ca}^A \end{aligned}\right\} \quad \sigma \text{ in ppm}$$

where $\sigma_{//}$ and $\sigma_{\perp}$ represent the parallel and perpendicular components of the screening tensor with respect to the molecular symmetry axis. Heavy nuclei with large chemical shift scales are obvious candidates for this kind of relaxation mechanism, the efficiency of which increases as the square of the applied magnetic field B_o. This effect has already been exhibited for ^{195}Pt for which a 20 Hz $^1J_{Pt-P}$ easily observed with an iron magnet, is not observed on reaching high field (5.9T).

The ^{195}Pt chemical shift anisotropy contribution to the observed line width (7) precludes the observation of $^1J_{Pt-P}$.

1.6.4 Scalar Relaxation Mechanism (sc)

When the observed A nucleus is J coupled with nucleus X, J_{A-X} value can be time-dependent because of a chemical exchange process that involves A and X and is defined by a life time τ_{ex}. If $1/T_{1sc}$ or $1/\tau_{ex} \ll J_{A-X}$, the coupling pattern will disappear and nucleus A will relax in a <u>scalar process of the first kind</u>.

In the extreme narrowing condition

$$\frac{1}{T_{1sc1}} = \frac{8}{3}\pi^2 J_{AX}^2 I_X(I_X + 1) \frac{\tau_{ex}}{1+4\pi^2(\nu_A-\nu_X)^2\tau_{ex}^2}$$

$$\frac{1}{T_{2sc1}} = \frac{4\pi^2}{3} J_{AX}^2 I_X(I_X + 1) (\tau_{ex} + \frac{\tau_{ex}}{1+4\pi^2(\nu_A-\nu_X)^2\tau_{ex}^2})$$

The efficiency of this mechanism is generally temperature- and concentration-dependent.

In a fast relaxation of nucleus X (e.g. ^{35}Cl, ^{81}Br, ^{14}N), but no breaking of the A-X bond, a scalar relaxation of the second kind occurs for A if $1/T_{2X} \gg J_{AX}$ and the corresponding T_{1sc2} and T_{2sc2} are obtained from the preceeding formulas with $\tau_{ex} = T_{2X}$.

1.6.5 Spin Rotation Relaxation Mechanism (sr)

In the extreme narrowing case, for a molecule A with cylindrical symmetry $T_{1sr}(T_{2sr})$ is expressed as

$$\frac{1}{T_{1sr}} = \frac{1}{T_{2sr}} = \frac{8\pi^2}{3}\frac{kT}{\hbar^2}(2I_\perp C_\perp^2 \tau_{sr\perp} + I_{//} C_{//}^2 \tau_{sr//})$$

where

$C_{//}$ and $C_\perp$ are the parallel and perpendicular components of the spin rotation coupling tensor expressed in Hz.

$I_\perp$ and $I_{//}$ are the components of the inertial moment along the symmetry axis.

$\tau_{sr\perp}$ and $\tau_{sr//}$ are the associated spin rotation correlation times.

Another formulation of $T_{1(2)sr}$ can be introduced by considering the components $b_{//}$ and $b_\perp$ of the fluctuating magnetic induction created at the nucleus site by the surrounding rotating charges (essentialy electrons) and expressed in Tesla (8,9). Then

$$\frac{1}{T_{1sr}} = \frac{1}{T_{2sr}} = \frac{8\pi^2}{3}\frac{kT}{\hbar^2}\gamma_A^2(2I_\perp b_\perp^2 \tau_{sr\perp} + I_{//} b_{//}^2 \tau_{sr//})$$

This equation shows T_{1sr} dependence on γ_A^2.

A similar dependence has recently been found for dimethyltellurium via ^{125}Te NMR (9). For spherically shaped molecules

$$C_\perp = C_{//} \quad \text{or} \quad b_\perp = b_{//},\ I_\perp = I_{//},\ \text{and}\ \tau_{sr\perp} = \tau_{sr//}$$

The spin-rotation mechanism can be large for small, rapidly reorientating functional groups, heavy spin 1/2 nuclei, and gas phases; contrary to all other relaxation mechanisms, T_{1sr} increases when the sample temperature decreases.

1.6.6 Quadrupolar Relaxation Mechanism (Q)

This mechanism is frequently encountered in heteronuclei NMR because 87 of the 116 magnetically active isotopes possess a spin value I > 1/2 and then a quadrupole moment Q.

In the extreme narrowing conditions

$$\frac{1}{T_{1Q}} = \frac{1}{T_{2Q}} = \frac{3\pi^2}{10}\frac{2I+3}{I^2(2I-1)}\left(1 + \frac{\eta_a^2}{3}\right)\left(\frac{e^2Qq}{h}\right)^2 \tau_c$$

Where τ_c is the reorientational correlation time of the A-X vector.

$$\eta_a = \frac{E_{xx} - E_{yy}}{E_{zz}} \quad \text{(asymmetry parameter)}$$

E_{xx}, E_{yy}, and E_{zz} define the principal components of the electrical tensor at the A nucleus, Q is the electric quadrupole moment of the A nucleus, e is the electric charge of the electron, and q is the electric field gradient along the A-X bond.

The quadrupole coupling constant (e^2Qq/h) in Hz, obtained from NMR (solid state), microwave (gas state), or nematic phase (liquid state) NMR measurements, can be found in the literature (10). This is an efficient mechanism; the observability (line width) of a given quadrupolar nucleus is strongly related to its quadrupole moment Q and the value of the electric field gradient q along the bond with which it is participating (high symmetry =sharp line). On the other hand, when several nuclei or isotopes on a similar electric gradient have quadrupole moments of the same order of magnitude but different nuclear spin I it should be remembered that the larger the I, the sharper the resonance line. This effect is due to the $(2I + 3)/I^2(2I - 1)$ factor in the T_{2Q} expression which is related to the line width by

$$\Delta\nu_{1/2} = \frac{1}{\pi T_{2Q}}$$

The table 1 lists the line widths ($\Delta\nu$) expected for spin I = 1,3/2, 2,..., 11/2 nuclei if a line width of 10 Hz is measured for the spin 1 nucleus and all other relevant nuclear parameters are identical.

Table 1

I	1	3/2	2	5/2	3	7/2	4	9/2	5	11/2
$\frac{2I + 3}{I^2(2I - 1)}$	5	1.333	0.583	0.320	0.200	0.163	0.098	0.074	0.058	0.046
$\Delta\nu$ Hz	10	2.6	1.16	0.64	0.4	0.32	0.19	0.14	0.11	0.09

CHAPTER 2

THE FOURIER TRANSFORM METHOD

All commercially available NMR spectrometers adapted for multinuclear observation operate in the Fourier mode. This recording procedure requires some understanding of the Fourier transform process. This chapter, then, emphasizes only the most important points to be kept in mind when engaging in Fourier transform NMR experiments on "other" nuclei. For more detailed developments on the NMR phenomenon itself the reader is referred to the following basic books or monographs: B1, B6, B9, B11, B18, B19, B20, B22, B25, B28, and B30.

2.1 THE ROTATING FRAME

In this representation the set of OXYZ axes (Figure 7a) precesses at a frequency equal to the Larmor precession of the resonant nucleus with respect to the fixed laboratory frame, Figure 7 illustrates the motion of the resonant nucleus magnetic moment $\vec{\mu}$ in this frame, off resonance (7b) and on (7c). It should be noted that on resonance is still valid when ν and ν_0 are slightly different (a few kHz); $\nu_0 = \omega_0/2\pi$ is the frequency of the applied RF field B_1 and $\nu = \omega/2\pi$ is the exact resonance frequency of the nucleus.

2.2 PULSE ANGLE IN THE ROTATING FRAME

When applied over a short time (τ sec) the Radio Frequency

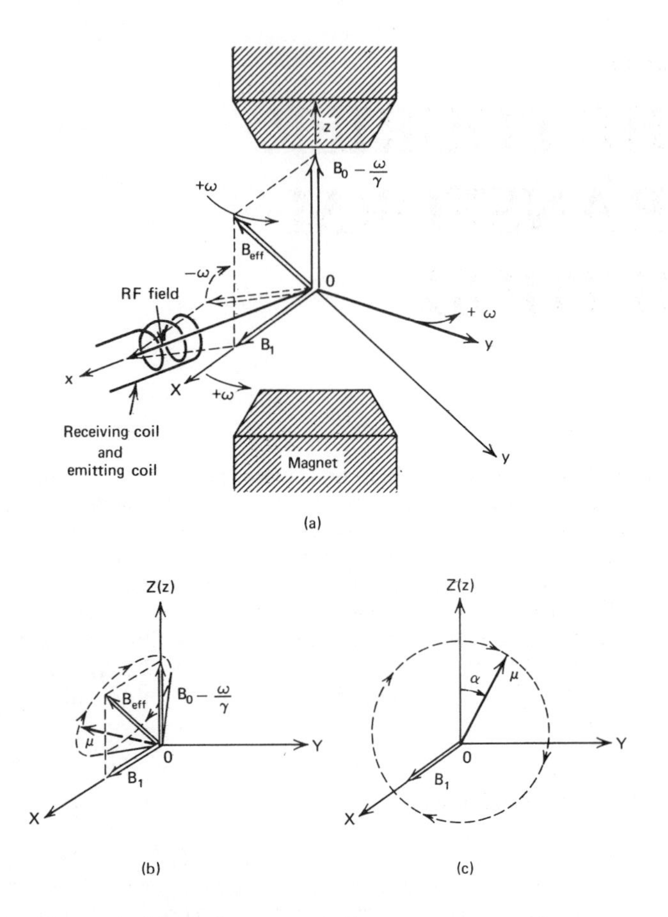

Figure 7. The rotating frame and the resonance phenomenon: (a). The different magnetic fields in the laboratory frame Oxyz and the rotating frame OXYZ; (b) motion of the nuclear magnetic moment μ in off resonance conditions (original position along OZ); (c) motion of μ in on-resonance condition in which $\omega = \omega_0 = \gamma B_0$ (original position along OZ; α, pulse angle)

field B_1 forces the magnetic moments to precess around it. As a result the macroscopic magnetization is tilted from its equilibrium position along B_o by a pulse angle α (Figure 7c)

$$\alpha = \frac{\omega_o}{2\pi} \tau = \frac{\gamma B_1}{2\pi} \tau \quad \text{(degrees)}$$

This pulse angle depends on the following points :

*The duration τ of the radio frequency pulse. When α = 90° the corresponding pulse is labeled 90° (τ_{90}). In some special pulse sequences (e.g., relaxation time measurements) the RF phase must be shifted sequentially. A subscript is then added to the corresponding pulse (90°x, 180°y,...).

*The amplitude of the radio frequency field B_1. This parameter depends on the output power delivered by the broadband emitter.On commercial spectrometers this power is nearly constant in the available frequency range; it can be decreased by inserting an attenuator between the emitter and the probehead; consequently all τ_{90} values are increased.

*The magnetogyric ratio. Because of the proportionality between α and γ, at constant B_1, the τ_{90} value for any A nucleus can be predicted, provided the τ_{90} for one nucleus has been determined with the same emitter, for instance,

$$\tau_{90}^{A} = \tau_{90}^{^{13}C} \cdot \frac{\gamma^{^{13}C}}{\gamma_A}$$

The determination of τ_{90} and τ_{180} is developed in Section 3.6.4.

2.3 THE FREE INDUCTION DECAY (FID)

2.3.1 The FID

After a pulse excitation a spin system tends to return to its equilibrium state and a free induction decay (FID) is detected by the spectrometer receiver coil (Figure 8). The decay rate is governed by an exponential term $\exp(-t/T_2^*)$, where

$$\frac{1}{T_2^*} = \frac{1}{T_{2s}} + \frac{1}{T_{2i}}$$

1. T_{2s} is the spin-spin relaxation time of the spin-system.
2. T_{2i} characterizes an instrumental spin-dephasing process due to magnetic-field inhomogeneities.
3. T_2^* is always shorter than T_1 because of the T_{2i} term, which means that the FID can disappear before the spin system reaches its thermodynamical equilibrium (T_1 process).

2.3.2 Time and Frequency Domains

The decay shown in Figure 8 cannot be directly interpreted by the chemist in terms of chemical shifts or coupling constants. It is well established (11; B1, B11, B17, B18, B19, B28, B30) that the Fourier transformation of this FID leads to the usual NMR spectrum and vice versa. This procedure is still valid when different FID are added or substracted.

2.4. THE DIGITALIZATION PROCESS

The frequency spectrum Fourier transformed from the FID is obtained with a dedicated computer coupled with a spectrometer; this process must be clearly understood, especially when dealing with the weak or widely shifted signals frequently encountered in heteronuclei NMR.

2.4.1 Sampling the FID

The receiving coil detects and delivers an analogic signal that must be digitized before being stored in the computer memory. This sampling operation is performed by an analog-to-digital converter (ADC; Figure 9). The actual FID voltage is measured every θ msec and the computer stores a number proportional to this voltage in each memory location.

The total sampling process duration AQT is obtained as

$$AQT = \theta N$$

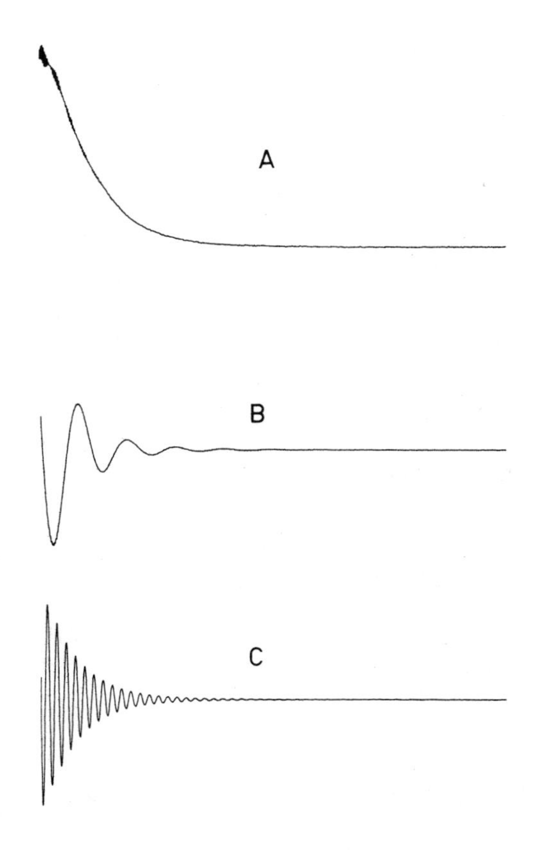

Figure 8. ^{199}Hg. FID of $HgMe_2$ (^1H broad band decoupled); (A) on resonance; (B) 10 Hz off resonance; (C) 50 Hz off resonance.

where N is the number of computer memory locations available for this operation and AQT is the acquisition time. The resolution of the digitized signal amplitude then depends on the number of bits the ADC is using for the sampling process. (Figure 10). As an example, to observe the ^{13}C satellites of $CHCl_3$ in the ^1H spectrum at least a 7-bit ADC is needed. Commercially available spectrometers are equipped with 12-to-16-bit ADCs.

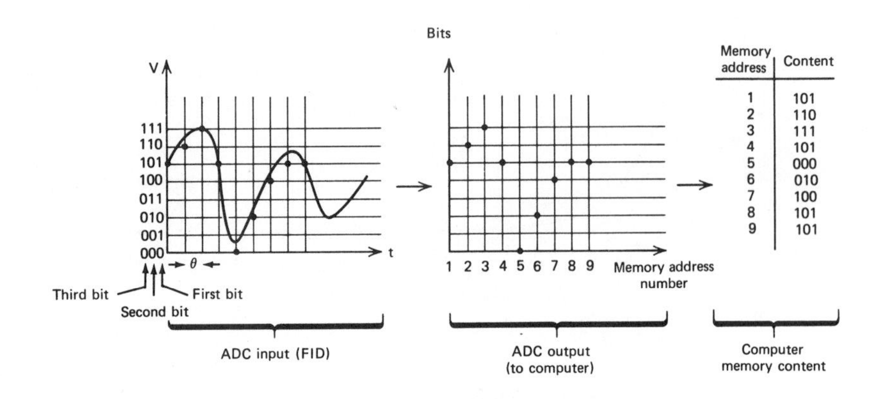

Figure 9. FID sampling process via a three bit ADC.

2.4.2 Sweep Width and Acquisition Time

The total number of memory locations (N) used during the sampling process and the acquisition time (AQT) are related to the observed frequency range (sweep width) by

$$\text{sweep width} = \frac{N}{2AQT} = SW \quad \text{(Nyquist theorem)}$$

2.4.3 Folding

This phenomenon, well known in FT NMR, is important when the spectra of unknow compounds of heavy elements with large chemical shift ranges are recorded. Two cases are encountered:

1. Sampling signal 1, Figure 11, which is outside the explored sweep width, will give the same result as sampling (hypothetical) signal 2, which is within the sweep-width frequency range. The transformed FID will produce a line whose frequency does not correspond to the real frequency of signal 1: instead,

   $$\nu^1_{app} = 2SW - \nu^1_{real}$$

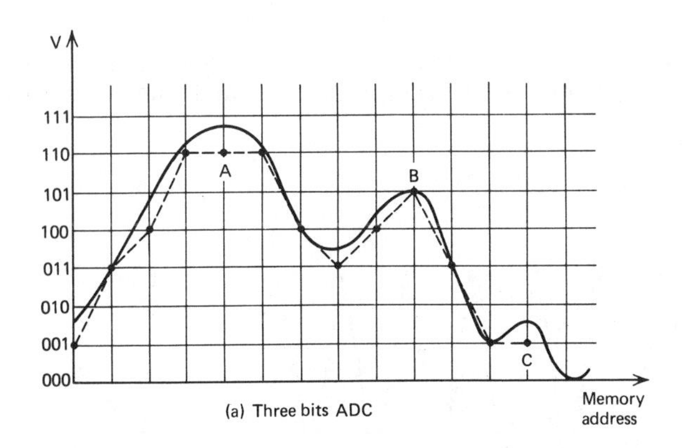

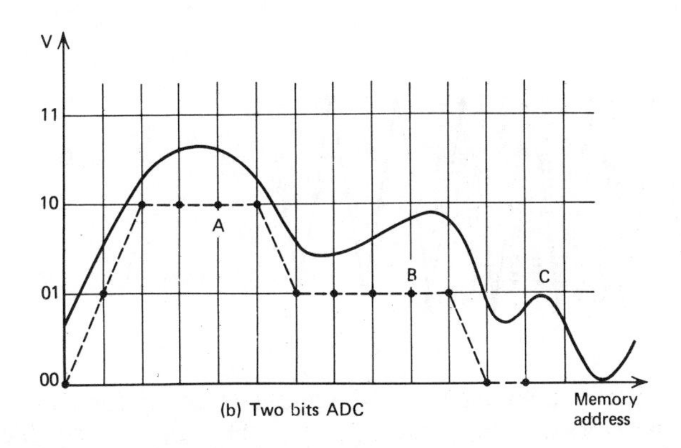

Figure 10. Influence of the number of bits of the ADC on the digitized overall signal. A and B signals are detected in both cases but in (a) signal C is digitized.

2. The detector cannot determine whether a resonance line is located at plus or minus $\delta\nu$ Hz from the carrier frequency ν_0. This gives rise, after the sampling process, to a folding back around ν_0, as presented in Figure 12 (arithmetic folding).

Folding recognition: By following Figure 11 and 12 it is easy to discriminate a folded line. If, for example, the carrier frequency offset is increased, keeping the sweep width constant, any real line will move toward to right side of the spectrum. A contrario, because of the symmetry effect around ν_0, any folded line will shift toward the left side of the spectrum.

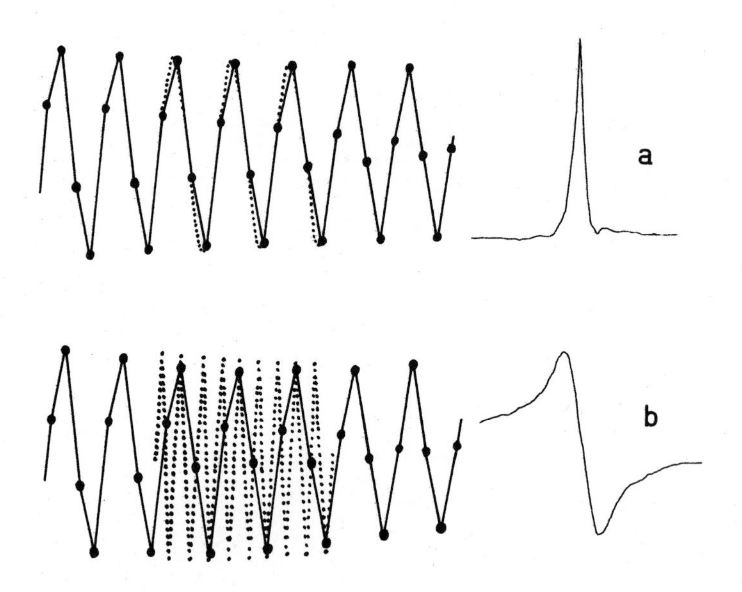

Figure 11. Folding phenomenon. ^1H spectrum of chloroform (SW = 100 Hz); (a) normal spectrum $\nu_0 - \nu = 50$ Hz; (b) folded line with $\nu_0 - \nu = 150$ Hz such as the folded line appears at the same frequency as in (a) (full line represents the theoretical FID, dotted line the real signal detected by the spectrometer; filled circles are data points).

2.4.4 Noise

The quality of a spectrum is a function of the signal-to-noise ratio S/N. Accumulating FIDs increases the signal as n and the random noise as $\sqrt{n}$, which corresponds to the $\sqrt{n}$ factor on S/N enhancement. When dealing with weak signals it is useless to let the spectrometer run for hours to gain in S/N; see the tabulation below in which the acquisition time is set to 1s and a S/N of 2 is assumed after 0.5 h.

S/N	Total experimental time (h)
4	2
8	8
16	32
32	128
64	512 (21.3 days!)

The condition becomes more favorable for short acquisition times (e.g., when observing quadrupolar nuclei with short T_1) in which rapid pulsing allows for overnight accumulation that can amount to a million or more scans.

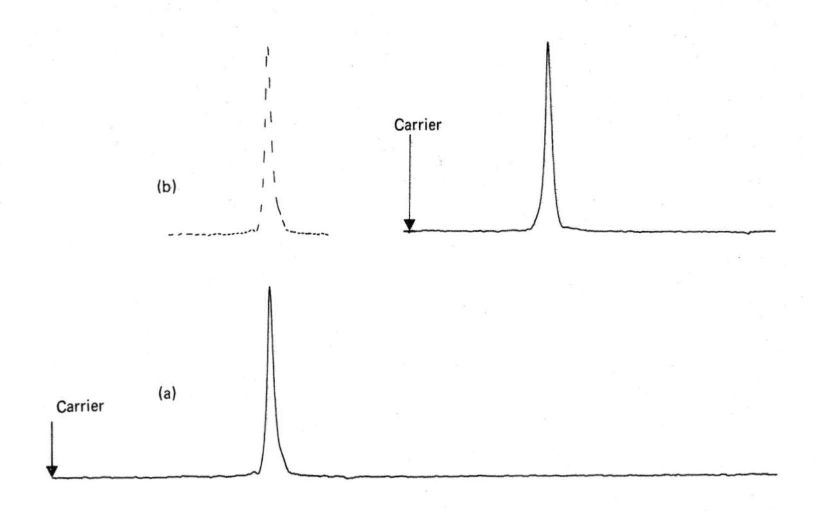

Figure 12. Arithmetic folding: (a) normal spectrum; (b) carrier frequency ν_o moved as $\nu_o^B = \nu_o^A - 250$ Hz (single detection mode).

2.4.5 Filtering, Quadrature Detection

The folding process during the sampling stage applies equally to the noise generated during the amplification of the signal before it reaches the ADC. This white noise is generally cut off by filtering and attenuating any signal outside the filter bandwith (Figure 13). It may be further reduced by using a quadrature detection scheme in which the carrier frequency is located at the center of the spectrum; the filter bandwith is set as 1.25 SW/2 and the analogic signal is routed to the ADCs in two separate channels, dephased one from the other by 90°. The resulting gain in S/N is then equal to $\sqrt{2}$, compared with the single detection mode, that divides the accumulation time by two.

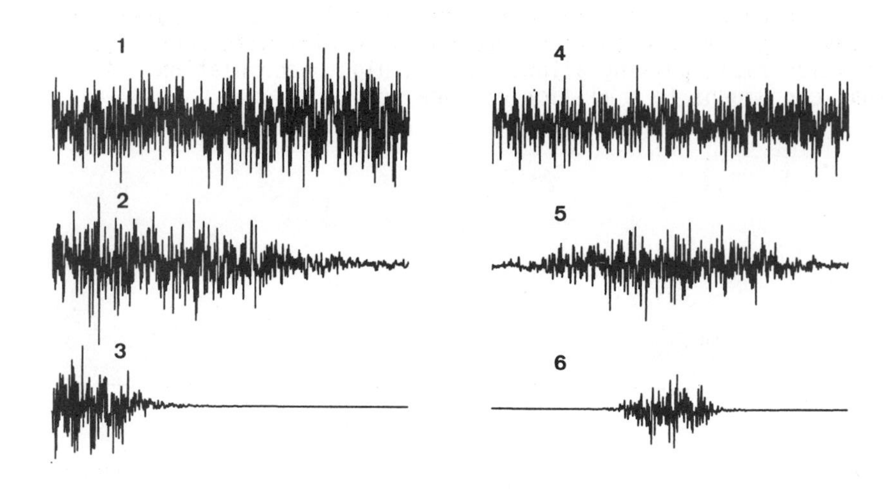

Figure 13. The noise in the frequency domain: SW, 25,000 Hz. Filter setting, 31,300 Hz (1,4), 15,000 Hz (2,5), 5000 Hz (3,6). Spectra 1, 2, and 3 are recorded in a single detection mode; spectra 4, 5, 6 in a quadrature detection mode. The $\sqrt{2}$ factor is clearly seen.

Remark. In this quadrature detection mode the folded lines do not obey the rules given in Chapter 2, and the single detection mode is preferred when a search for an unknown resonance is made.

2.5 PRECISION: RESOLUTION

2.5.1 The Detection Threshold

An analogic signal is detected by the ADC if its corresponding voltage is greater than the minimum voltage necessary to excite the ADC's first bit. This limit is, in fact, lowered by the digitized noise as shown in Figure 14.

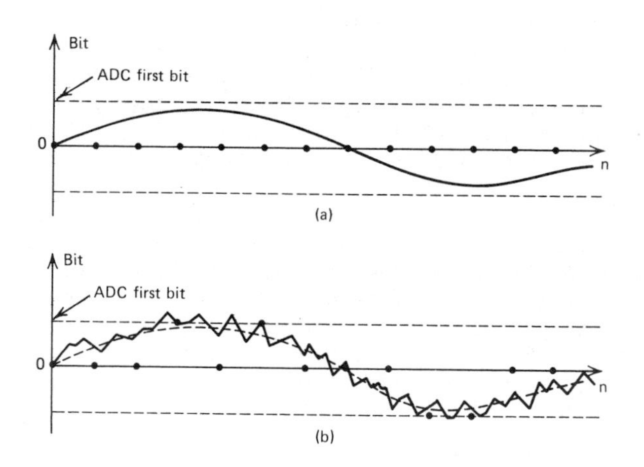

Figure 14. Influence of the noise on signal detection: (a) signal without noise; (b) signal with noise. Filled circles represent data stored in computer memory.

2.5.2 Saturation

On the contrary, if the sampled signal overloads the ADC's capabilities, it leads to a saturation effect on the computer's memory ("clipped" signal), and the transformed line shows strong distorsions (Figure 15). A careful adjustment of the spectrometer gain will allow the experimentalist to determine the optimum recording conditions.

2.5.3 Precision On The Resonance Frequency; Line Shape

If the decay is stored in N memory locations, the resulting usable frequency spectrum will be displayed over N/2 points. The frequency resolution for this experiment is

$$d = \frac{2SW}{N} \text{ Hz}$$

and the line position is obtained with a ±d Hz accuracy. For spin 1/2 heavy nuclei (^{57}Fe, ^{103}Rh, ^{183}W,...) with rather large chemical shift ranges, which imply large SW and sharp lines (1 Hz), the situation can become critical because N, the number of data points available, is generally limited. As an example, a 25,000-Hz sweep width associated with a 16 K memory block produces a resolution of 3.0 Hz; therefore a resonance can be missed if its line width is smaller than 3 Hz. It is far better, then, to reduce the sweep width to 5000Hz and to move the carrier frequency by 5000 Hz steps after completion of the required number of scans and the Fourier transformation in order to explore the entire 25,000-Hz spectral region. This procedure will hold equally well in line-shape analysis or precise quantitative studies; for example in ^{27}Al NMR for which broad, sharp lines can be detected and in which each line should be defined by 10 or more memory points (Figure 16a, 16b).

2.6 DISPLAY OF THE NMR SPECTRUM

2.6.1 Normal Mode or Absorption Mode

$$V = \frac{A}{1 + 4\pi^2(\nu_0 - \nu)^2 T_2^{*2}}$$

Produces a Lorentzian line shape (Figure 17b)

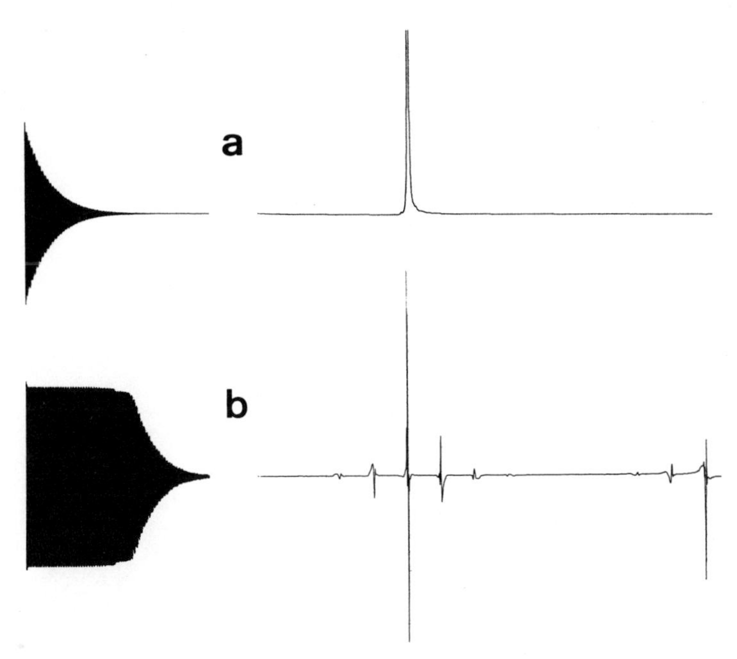

Figure 15. ^{27}Al spectrum of $AlCl_4^-$: (a) normal spectrum; (b) ADC overloaded and corresponding spectrum.

2.6.2 Dispersion Mode

$$u = \frac{A'(\nu_o - \nu)}{1 + 4\pi 2\,(\nu_o - \nu)^2\, T_2^{*2}}$$

This mode may be used to detect broad lines (Figure 17a).

2.6.3 Magnitude Mode

$$m_{(\nu)} = \sqrt{u^2 + v^2}$$

In this mode the line integral fails to converge and phase information is lost (Figure 17c).

2.6.4 Power Mode

$$P_{(\nu)} = (u^2 + v^2)$$

This mode still provides a Lorentzian line shape, but the integral is proportionnal to the square of the number of resonating nuclei (to be used when phase correction is difficult or not needed) in Figure 17d.

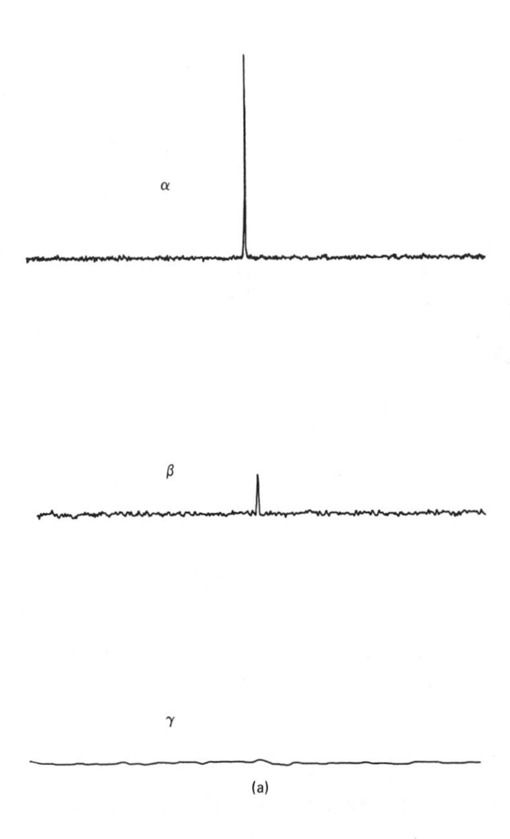

Figure 16a. Influence of the number of data points on a, the detection of a sharp signal (^{95}Mo spectrum of MoO_4^{2-}); α, 0,6 Hz/pt; β, 10 Hz/pt; γ, 200 Hz/pt.

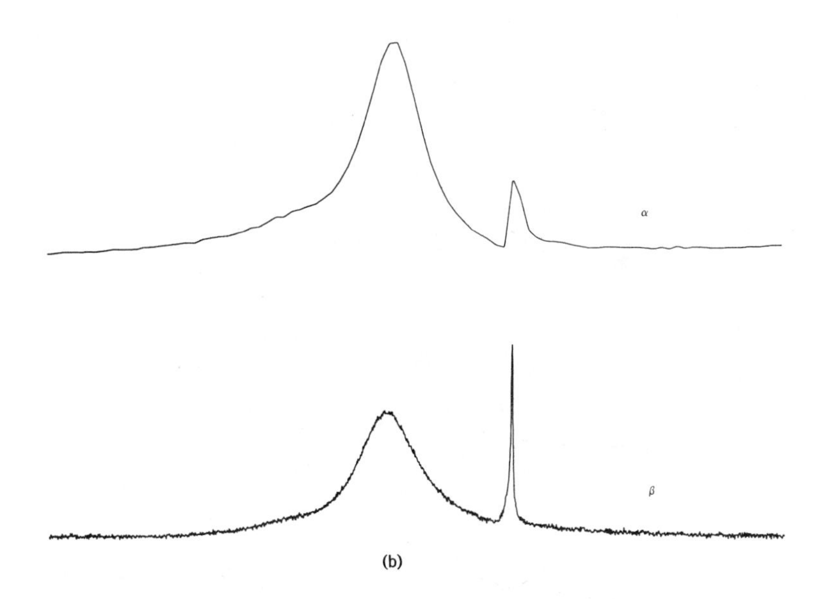

Figure 16b. The precision of a quantitative measurement ($AlCl_3$ in CH_3CN, low field region, ^{27}Al spectrum, SW = 15,000 Hz, same experimental recording conditions); α, 1 K memory; β, 32 K memory.

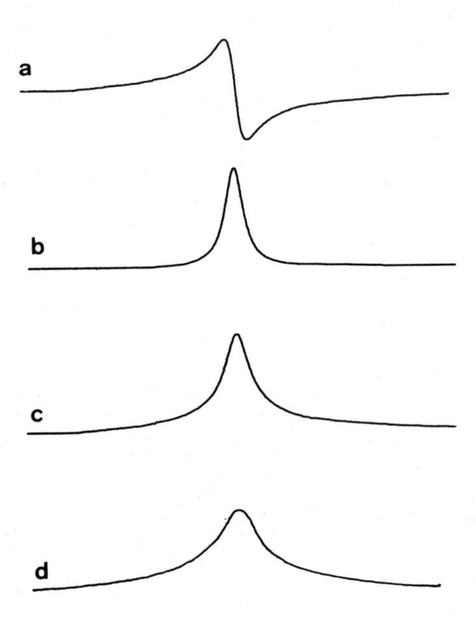

Figure 17. ^{71}Ga spectrum: (a) u mode; (b) v mode; (c) power spectrum; (d) magnitude spectrum.

CHAPTER 3

SPECTRUM RECORDING

Modern spectrometers permit the full use of the efficient pulse sequences or FID mathematical treatments recently developed. Before detecting and plotting the resonance, however, it is sometimes useful to check the experimental set-up. These points are reviewed in this chapter with literature coverage up to mid-1980.

3.1 SAMPLE PREPARATION

3.1.1 General Case

The following hints are worth remembering:

1. Avoid any solvent for which resonance rises in the region in which the solute resonances are expected.
2. Prepare a solution as concentrated as possible, the height of which in the sample tube will be slightly higher than that of the receiving coil.
3. Field frequency locking (cf. Section 3.2) will be achieved (generally) by solvent deuterium resonance. A reference compound (cf. Section 3.3) can be added to the solution.
4. The solution viscosity should be as low as possible, especially for quadrupolar nuclei, to avoid severe line broadening.

3.1.2 Sample Preparation for Quantitative or Dynamic Measurements (NOE, T_1 and T_2, Line-Shape Fitting)

1. Filter out any solid particules left in the solution.
2. Remove the oxygen dissolved in the solvent (paramagnetic contribution) with three to four freeze-thaw cycles under high vacuum and sample tube sealing, or flush the solution with rare gas (Argon) for 10 min (sufficient for NOE, T_1, and T_2 measurements when $T_{1(2)} < 50$ s), or use (with care) special oxygen removing reagent (12).

3.1.3 Special Cases

Gases are, in general, sealed under pressure in thick-walled NMR tubes. When studying anisotropic phases (nematics, smectics, etc.) degas the sample and check its anisotropic phase temperature range. Iron magnets require that the sample tube be left in the magnet gap for 1 or 2 h before starting the experiment which is then completed without spinning the tube. For superconducting systems the colinearity of the magnetic field and tube axis allows sample spinning and thus better resolution.

3.2 THE LOCK SYSTEM

This accessory, now common in all FT spectrometers, maintains the B_o/ν_o ratio constant by monitoring the frequency of a given NMR line; any detected drift in this resonance position is fed back by an automatic control loop as a correcting voltage to the magnet power supply to achieve the correct field value.

3.2.1 Homonuclear, Heteronuclear Lock

This lock is called homonuclear when the locking resonance is in the frequency range of the observed nucleus; heteronuclear when the locking resonance falls within a frequency range far from the resonance frequency of the nucleus under study. Deuterium is the most preferred locking isotope because of the availability of deuterated solvents. Proton or fluorine is sometimes used when proton broad-band decoupling is not required.

3.2.2 Internal, External Lock

Three situations are encountered:

1. Internal lock in which the locking nucleus is furnished by the solvent (deuterium, proton, or fluorine);

2. External lock of the first kind in which the locking substance is contained in a capillary tube that slides into the NMR tube or surrounds the NMR sample tube (e.g., an 8-mm tube in a 10-mm tube filled with the locking substance);

3. External lock of the second kind in which a small ampoule that contains the locking substance is permanently located near the sample tube in the probehead. A small coil wound around this vial detects the lock signal. This attachment cannot be used for high resolution or T_2 measurements (13) because of the completely different field homogeneity in the sample and lock-substance positions.

3.2.3 Lock Tuning

Because the lock detection is performed in a continuous wave mode care must be taken to avoid lock-signal saturation, especially when compounds with long T_1 values are used (e.g., acetone d_6 for a deuterium lock, C_6F_6, $CFCl_3$ for a fluorine lock). When saturation occurs, the lock becomes unstable because of the variation in the amplitude of the lock resonance. On the other hand, because the lock-detecting coil is tuned on a fixed frequency, generally centered in the middle of the frequency range of the locking nucleus, a field-sweep facility is always provided to match the B_0 value with the exact resonance frequency of the locking substance. This operation will prove important when different lock samples are used for example, $CDCl_3$, C_6D_6, CD_3-CO-CD_3 (cf. Section 3.4) or when the sample lock frequency depends on the volumic susceptibility χ_v of the solution (cf. Section 3.3), the temperature, or the concentration.

3.3 REFERENCE

A specific frequency range, therefore a specific reference, corresponds to all detectable isotopes. Thus it is most important in multinuclear NMR to define all measured chemical shifts precisely (B10, Vol.3, p. 221; Chapter 1).

3.3.1 Types of Reference

Internal. This is an inert substance, that is added to the solution to provide a unique single peak for referencing purposes; for example,

Nitromethane for ^{15}N
Water for ^{17}O
TMS for ^{29}Si
Sodium tungstate for ^{183}W

In any case, a clear description of the reference (molarity, pH of the solution) will help in a comparison of experiments.
External. In heteronuclear NMR it is often impossible to add the reference compound directly to the solution under study. In the procedure to adopt the reference substance is added to a capillary inserted in the NMR sample tube.
Substitution. Some samples cannot accept a capillary tube. After the sample spectrum is recorded the tubes are exchanged and the reference spectrum is recorded and plotted with the same parameters. On the other hand, most modern FT spectrometers allow the reference frequency to be stored in the computer memory; this produces a fast spectrum calibration.
Secondary. If the frequency of the reference line is far outside the observed spectral region (which is common in heteronuclei NMR), an appropriate compound or a solvent peak with a known chemical shift can be used for secondary referencing: $Mo(CO)_6$ in molybdenum NMR resonates at -1858 ppm from MoO_4^{2-} (reference), whereas its sharp resonance is closed to $Mo(CO)_5L$ type resonances; XeF^+ has been used as a secondary reference in ^{129}Xe NMR in which $XeOF_4$ is accepted. This secondary referencing diminishes the chemical shift measurement accuracy by a factor of 2.

3.3.2 Influence of χ_v

When measured by an internal reference the chemical shifts do not depend on the volumic susceptibility χ_v of the solution.

When external referencing is used (cf. the preceding section) a correction must be applied. Then for an iron magnet

$$\delta_{ppm} = \delta_{meas} + \frac{2\pi}{3} (\chi_{ref} - \chi_{sample}) 10^6$$

For a superconducting magnet

$$\delta_{ppm} = \delta_{meas} - \frac{4\pi}{3} (\chi_{ref} - \chi_{sample}) 10^6$$

where χ_{ref} and χ_{sample} represent the volumic susceptibilities of the reference and sample solutions respectively. These formulas show clearly the chemical shift discrepancy that may develop for the same compound, externally referenced, when the measurement is carried out with an iron magnet spectrometer, $\delta_{meas\perp}$ (B_0 perpendicular to the sample tube axis) or a superconducting, $\delta_{meas//}$ (B_0 parallel to the sample tube axis) if no χ_v correction is applied (14,15). This difference is

$$\delta_{\perp} - \delta_{//} = 2\pi (\chi_{sample} - \chi_{ref}) 10^6$$

which for example amounts to 0.68 ppm for the ^{15}N chemical shift of pyridine dissolved in benzene and referenced with respect to external aqueous NO_3^-.

3.3.3 In General

The frequency of the reference line can be solvent or temperature-dependent. For heavy elements the temperature coefficient on the chemical shift may be as high as 3.5 ppm/°C(^{59}Co). When these nuclei are dealt with it must be made certain that thermal equilibrium has been reached in the probe before measurement is started; attention must also be paid to possible temperature gradients arising from pulse trains, decoupler power, etc. For the purpose of future comparison the temperature at which the experiment is performed must be indicated.

3.4. CHEMICAL SHIFT MEASUREMENTS

Because the first goal of many NMR experiments is a chemical shift measurement, we stress the importance of correct data manipulation and presentation. The upsurge of increasing NMR data on many elements of the periodic table demands a clear statement of all experimental details in published data tables, if only to allow full use of earlier work!

A simple guideline should be to describe precisely

(a) the mode of referencing (internal, external);
(b) the physical state of reference and solute (molarity, solvent);
(c) the temperature at which the experiment has been carried out;
(d) the lock substance used;
(c) the chemical shift or coupling constant accuracy.

3.4.1 Internal Referencing

The chemical shift is directly measured from the position of the sample line with respect to the reference line. Even in this simple case solvent effects can be difficult to sort out (B10, Vol.3, p.231).

3.4.2 External Referencing

Any sample-reference, solute-solute interaction is nulled. A normalized reference is worth using. As an example,

CH_3NO_2 neat for ^{14}N, ^{15}N

$Al(H_2O)_6^{3+}$ 1M in H_2O for ^{27}Al

$Te(CH_3)_2$ as pure liquid for ^{123}Te, ^{125}Te

A χ_v correction must be applied as previously indicated (Section 3.3.2).

3.4.3 Substitution Referencing

The normalized reference is dissolved in a solvent used for locking. The exact resonance frequency of this reference is then measured with, when necessary, its temperature dependence. Two situations can develop

Sample and Reference Lock Substances are Identical. The chemical shift of the sample is given by

$$\delta_s = \frac{\nu_s - \nu_o}{\nu_o} \text{ (in ppm)}$$

where $\nu_{o(s)}$ is the resonance frequency of the reference (sample). This operating procedure avoids a χ_v correction, and the use of a well-defined referencing compound allows exact comparison of experiments.

Sample and Reference Lock Substances are Different. Because of the fixed frequency circuitry used to lock the spectrometer the B_o field must be slightly modified to match the resonance condition for the lock substance of sample and reference, in which case a correction must be applied, given by

$$\delta_{sample} = (\delta_{sample\ (measured)} + \Delta)\ ppm$$

where

$$\Delta = (\delta^{sample}_{lock} - \delta^{ref}_{lock})$$

is the difference in chemical shift, expressed in ppm, between the two locking resonances. If a deuterated lock substance is used as solvent, then neglecting solvent effects, the corresponding δ lock values (see Table 2) are safe:

As an example, when ^{77}Se at 2.1 T is measured

$\nu Se(CH_3)_2$ (reference): 17.168816 Hz, lock $CDCl_3$

$\nu({}_pCH_3\text{-O-}\emptyset)_2Se_2$ (sample): 17.177485 Hz, lock $(CD_3)_2CO$

δ sample (measured): 504.9 ppm

δ sample (corrected): 504.9 ppm + (2.17 - 7.27) = 499.8 ppm

External Lock of the Second Kind. No lock chemical shift correction is required, but nevertheless a χ_v correction must be performed, as explained in Section 3.3.2.

Note. In any case, the lock chemical shift correction can be important in contrast to the χ_v correction.

3.4.4 Universal Referencing

Several groups (16,17,18,19; B21,p.135) have recommended that all measured resonance frequencies be related to a field B_o in which the protons of TMS resonate at exactly 100 MHz. The symbol chosen to define these frequencies is Ξ.

Table 2

Solvent	Lock (ppm)
TMS d_6	0
C_6D_{12}	1.43
CD_3CN	2.00
CD_3I	2.15
$(CD_3)_2CO$	2.17
$(CD_3)_2SO$	2.62
p.dioxane d_4	3.70
CD_3NO_2	4.33
CD_2Cl_2	5.30
$CDCl_3$	7.27
C_6D_6	7.37
Toluene	7.17
d_8	2.32

This presentation is, in fact, handy in multinuclear NMR experiments; it has been adopted for the second part of this book, with all necessary information for the calculation of the corresponding frequency on a given spectrometer. On the other hand, these Ξ values are somewhat difficult to manipulate, especially when relative δ measurements are dealt with. The substitution referencing method described in Section 3.4.3 is considered the simplest to handle because only one determination of the reference frequency is required.

3.5 χ_v MEASUREMENTS

It is not necessary to measure this parameter on the nucleus under study; the most sensitive nucleus present in the solution (^1H, ^{19}F, ^{31}P,...) can be used.

Note. For all the following methods a careful study of the original papers is strongly recommended for setting up the experimental details.

3.5.1 χ_v Tables

This is, indeed, a trivial method which has nevertheless proved to be sufficient in many cases. A wealth of χ_v values can be found in the following publications: Handbook of Chemistry and Physics (Section E), (B9, Vol.1, Appendix C).

3.5.2 Sphere and Cylinder Method (20)

Two different chemical shifts can be measured for a given compound when the substance is contained in a sphere topped with a capillary. The difference is expressed as

$$\delta_{cylinder} - \delta_{sphere} = k(\chi_{ref} - \chi_{sol})\ 10^6$$

where k = $2\pi/3$ for an ideal sphere and cylinder; its real value, which corresponds to the experimentl set-up, is calculated from compounds with known χ_v.

3.5.3 Capillary Method (21,22)

A substance with a known χ_v is contained in a capillary which is installed in the NMR sample tube. This is a delicate method because of the critical shim adjustement required under nonspinning conditions (22); some modifications have been proposed (17,23,24,25).

3.5.4 Adams' Method (26)

Two concentric tubes are used. The inner tube is filled with a reference compound whose spectrum is recorded. The annular space is then filled with the solution of unknown χ_v and the reference line position is recorded again. Then

$$\chi_{v(sol)} = \chi_{v(ref)} \frac{\delta_{ref}^{solution}}{\delta_{ref}^{alone}}$$

This method requires a calibration curve.

3.5.5 Becconsal's Method (14)

This method is based on the determination of the chemical shift difference of a chosen line of the solution when measured successively with an iron magnet and a superconducting magnet; χ_v value determination is then straightforward (cf. Section 3.3.2).

3.6 SPECTRUM RECORDING

After the spectrometer is tuned on a reference line for a given nucleus,spectra of unknown compounds must be recorded. This procedure requires certain general rules be followed.

3.6.1 Carrier Offset

This parameter and the sweep width (SW) are of prime importance because they define the spectral region under measurement in which the resonances are expected. For heteronuclei a good guide for chemical shift ranges can be the oxydation state of the nucleus. Part 2 of this book will help the beginner on that point. For a first trial,a spectral width of ca. 10,000 Hz, a 45° pulse angle, and a repetition rate of 1 sec can be chosen. Once the resonance(s) is (are) detected, it is mandatory to determine whether the line(s) is (are) folded (see Section 2.4). If single detection is used, the displacement of the carrier toward high frequency must result in the movement of all lines by the same amount (in Hz) with respect to the carrier. If not, the offset must be recalculated or the spectral width increased to record the entire spectrum properly. If quadrature detection is used and folding is suspected, it is easier to return to the single detection mode and to define offset and spectral width accordingly. In any case, because of digital filtering, a line away from the carrier frequency of ±(3 SW) Hz in single detection or ±(1.5 SW) Hz in quadrature detection will hardly be detected.

3.6.2 Sweep Width and Memory Size

The sweep width is generally chosen to match the entire spectrum. For a given sweep width the memory size sets the acquisition time value, hence the spectral resolution. Because at least one point is needed to define a resonance line, the spectral resolution should be such that

$$\Delta_{\nu 1/2} > \frac{2\ \mathrm{SW}}{\text{number of points for acquisition}} \quad \text{in Hz}$$

As a practical hint sharp lines associated with rather long T_1 will need good resolution (long acquisition times); broad lines (short T_1) are acquired under low resolution conditions. In general, the acquisition time should approximate $3T_2^*$.

3.6.3 Pulse Interval, Pulse Angle

After 10 pulses or so the spin system reaches a steady state that actually determines the strength of the signal to be acquired. Optimal pulse angle α_0 and pulse interval t_w are related to the rate of recovery of the magnetization T_1 via (11):

$$\cos \alpha_0 = \exp\left(-\frac{t_w}{T_1}\right) \quad \text{(Figure 18)}$$

In a spectrum in which several lines with different T_1 are present a compromise must be found. Long T_1s require long t_w or a small pulse angle. The T_1 values can be reduced by adding relaxagents such as chromium acetyl-acetonate or cryptates (27) for aqueous solutions, provided that these compounds do not react with the solute. For intensity measurements a pulse angle of ca. 30° is recommended to level out the problems that could develop from non uniform power distribution across the explored spectral width due to finite pulse length. In the measurement of time-dependent systems (CIDNP experiments, chemical exchange process, etc.) an important recording condition, often ignored, has been stressed by Ernst et al. (28) who showed unambiguously the importance of pulse angles shorter than 15° for these experiments.

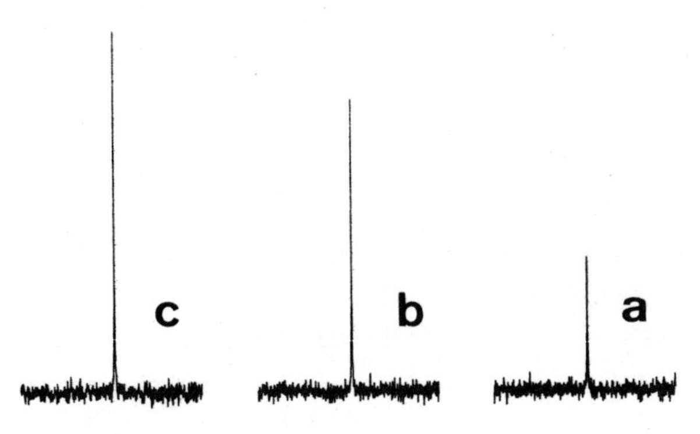

Figure 18. Influence of pulse angle value on a given ^{15}N signal. Sample $Na^{15}NO_3$, M/D_2O, waiting time 5s, 100 pulses: (a) $\alpha = 90°$, (b) $\alpha = 45°$; (c) $\alpha = 15°$.

3.6.4 Determination of a 90° Pulse Angle

Because of the cosine dependance of the signal strength on pulse angle value (flat maximum), it is generally more convenient to determine the 180° pulse that corresponds to a nulled signal; the 90° value is then half the value found. In this procedure a waiting time of at least $5T_1$ of the measured line is required. A simple procedure for this experiment consists in varying the pulse value τ_i and recording the corresponding i^{th} spectrum. A straightforward interpolation between the peak intensities around the first minimum yields τ_{180} (Figure 19). A precise determination of a 90° pulse angle requires careful adjustment of the liquid height in the sample tube to avoid diffusion effects or exciting radio frequency field inhomogeneity.

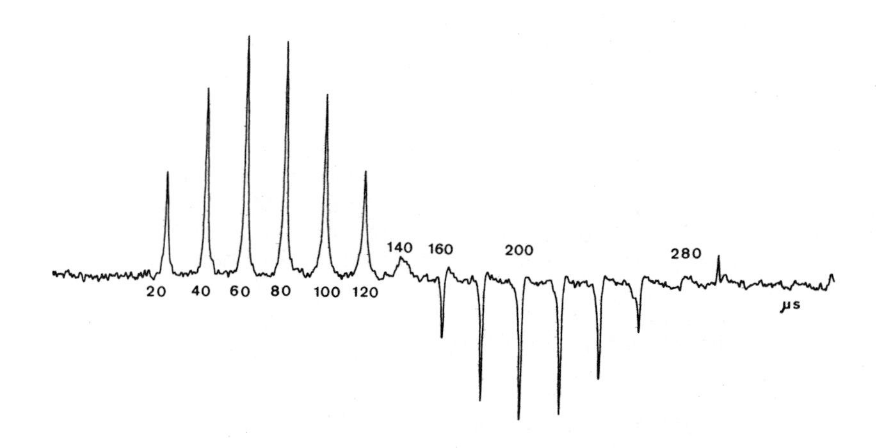

Figure 19. τ_{90} determination for ^{95}Mo (sample MoO_4^{2-}): τ_{90} = 73 μs, τ_{180} = 146 μs.

When dealing with low sensitivity nuclei for which a reasonable signal over noise would be reached only after prohibitive spectrometer time the 90° pulse angle value can be approximated, provided that a nucleus with high sensitivity resonates in the same frequency range, in which case the τ_{90} pulses are related by the inverse ratio of respective magnetogyric ratios. As an example, if $\tau_{90}^{^{35}Cl}$ is 17 μs, the τ_{90} for ^{15}N will be close to

$$17 . \frac{\gamma_{^{35}Cl}}{\gamma_{^{15}N}} \sim 16 \text{ μs}$$

On the other hand, τ_{90} values for one nucleus vary with the ionic strength of the solution (93), especially at high observing frequencies.

3.6.5 Spectrometer Gain

Before the accumulations are started the receiver gain must be optimized. Two situations apply:

1. Weak of very weak signals will be detected only if sufficient noise is present to excite the first bit of the ADC, thus allowing coherent signal accumulation (cf. Section 2.5.1, Figure 14).

2. On the contrary, with strong signals "clipping" by the ADC must be avoided; otherwise saturation will produce severe line distortion (cf Section 2.5.2, Figure 15). Precise adjustement of this operating condition will generally be found in the manufacturer's manual.

3.6.6 Delay Before Acquisition

To avoid pulse breakthrough or acoustic ringing, two enhanced problems when observing low γ nuclei, a delay must be occasioned between the end of the pulse and the beginning of the acquisition. Neglect of this delay will result in "rolling" base lines that will render the transformed spectrum useless. On the other hand, the delay introduces an extra dephasing process that is generally not considered in the phase correction subroutines of the spectrometer. A more severe drawback of this unavoidable delay at low frequencies is the loss of signal information for broad lines with associated short T_2.

A hopeless situation must then be faced when high resolution NMR (liquid state, weak signal) must be done under solid-state NMR conditions (very broad lines, very short T_2), especially when quadrupolar nuclei such as covalently bound sulfur 33, niobium 93, and chlorine 35, are under study for which total NMR information is present in the very first points of the FID. No acceptable solutions have yet been found for this problem. Intermediate situations can be tackled in software based FID manipulations (cf Section 3.7.1).

3.7 FID AND SPECTRUM MANIPULATIONS

All commercially available spectrometers offer a software package that accommodates a number of FID or spectrum manipulations. These operations can prove useful, provided there is evidence of their exact consequences in the final spectrum.

3.7.1 The FID

Before Fourier transformation of the FID some operations can be done in memory to avoid base-line distortion and to increase the effective resolution, or the apparent signal-over-noise ratio.

Base Line Distortion. This distortion comes from the pulse breakthrough or from probe acoustic ringing, especially at low frequencies with the result that there is saturation of the first 10 to 50 memory words. A flat baseline can then be obtained by:

* Left shifting the FID. This procedure is equivalent to a delay before the acquisition and introduces a third-order dephasing (Figure 20).

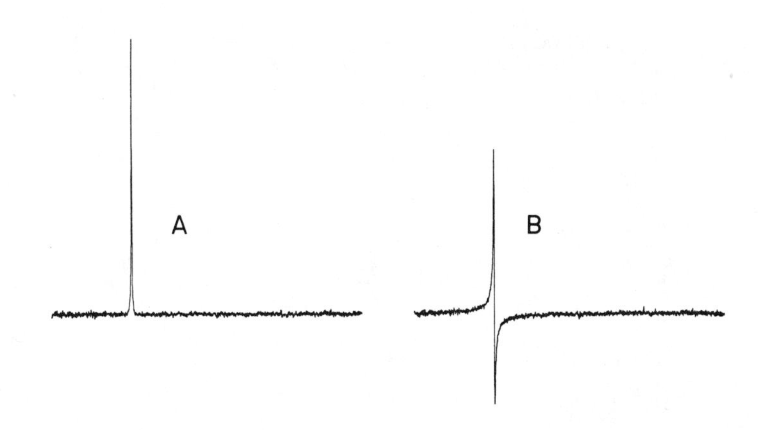

Figure 20. The effect of left shifting the FID: (A) normal spectrum; (B) the same phase corrections as in (A) but with a shift of the FID 10 points to the left before Fourier transformation.

* Zero filling the beginning of the FID, an operation that stresses the (sin x/x) appearance of the peak (Figure 21).

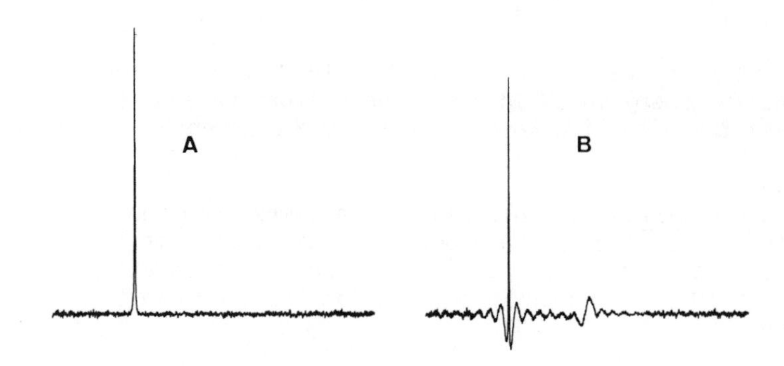

Figure 21. The effect of zero-filling the first points of an FID: (A) normal spectrum; (B) the same phase corrections as in (A) but the first 60 points of the FID are zeroed.

✱ Applying a trapezoidal window to replace the left shift procedure and to reduce any truncations effects (Figure 22) due to an acquisition time that is too small for T_2^*.

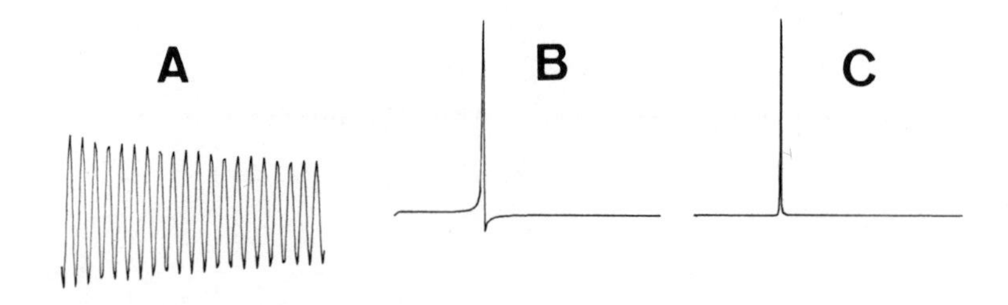

Figure 22. Truncation effect: (A) the FID; (B) transformed spectrum; (C) normal spectrum.

Sensitivity Enhancement. This is done by multiplying each point of the FID with a decaying exponential, exp(-t/a) whose positive time constant a matches the apparent FID T_2^*.

This convolution process introduces line broadening of $-1/\pi a$ and the time constant of the exponential is generally typed in as a broadening factor (in Hz).

Resolution Enhancement. All the methods used to enhance the resolution need a good signal-to-noise ratio to start with because they generally decrease the signal intensity. Several procedures have been proposed:

* Multiply the FID by a weighting exponential function, $\exp(-t/a)$, with $a < 0$. (B31, Vol.2, p.1).
* Multiply the FID by a weighting sine function (29, 31).
* Shift the dispersion mode spectrum by two to three points to the left and substract it from be initial spectrum (30).
* Multiply the FID by a weighting binomial exponential function (32, 33, 66), $\exp(-at-bt^2)$.

The interested reader may also refer to other proposed convolution procedures (34, 35, 36).

3.7.2 The Spectrum

Integral. This parameter is critical in a FT spectrometer because it is sensitive to DC bias or phase misadjustment in the transformed spectrum; the result would be "tilted" intregral lines. The computer generally defines the noise level by summing and weighting the address content of several blank points; a peak will then be detected and integrated if its address contents overtake the defined noise level. Standard FT programs allow the spectrum to be integrated stepwise by most efficient phase or DC bias adjustment.

Peak Position. The dedicated computer associated with the spectrometer readily provides this information. The peak-picking subroutine will print out the peak frequency and chemical shift with respect to any chosen reference line, provided that the peak's most important address content overrides a preset threshold. It must be remembered that the precision on the peak position is not defined by the values listed by the computer but by the digital resolution (cf. Section 2.5.3).

3.8 DECOUPLING PROCEDURES

Heteronuclei NMR sometimes requires decoupling experiments. The versatility of modern multinuclei FT spectrometers will certainly help to develop new decoupling schemes or new homo or hetero decoupling experiments on isotopes other than proton or fluorine.

3.8.1 Heteronuclear Decoupling

This decoupling experiment consists in observing a given nucleus while irradiating another in a different spectral region. Several hetero decoupling states can be selected:

* Broadband. The entire chemical shift range of the irradiated nucleus is swept by a modulation scheme in the decoupling frequency. In a chemical shift range that is too large (^{19}F, ^{31}P,...) and demands far too much power to be fed in the probehead, it must be broad-band decoupled stepwise. Broadband decoupling will generally require some attention to temperature gradients in the sample, especially for heavy nuclei with important temperature dependent chemical shifts.

* Selective. A specific line can be irradiated continuously under low power or saturated by a 180° selective decoupling pulse, followed by a 90° nonselective observing pulse; the last described experiment is known as selective population inversion (SPI) (37, 39) or selective population transfer (SPT) (38) and allows study of exchange (40) or relaxation (41, B31, Vol. 9, p.190) rates, NOE effects, or the sign of indirect (38, 42, 43) or direct (44) coupling constants.

* Broad-band or selective off resonance. Offsetting the decoupling frequency a few hundred herz away from either end of the spectral range results in a partially decoupled observed spectrum. Such an "off resonance" pattern can be useful in the assignment of different structural moieties in the molecule (B20).

3.8.2 Homonuclear Decoupling

In these experiments the observed and decoupled nuclei are of the same type. This decoupling mode requires that the decoupler be switched on during the acquisition process, in time sharing with the receiver, to avoid preamplifier saturation (45). A small frequency shift (Bloch-Siegert) is noticeable in resonances that approach the irradiation frequency (46).

3.8.3 INDOR

INDOR has been identified only as a continuous wave experiment but is now routinely performed on FT spectrometers:

* The double resonance difference spectroscopy (DRDS) (47) method substracts a selectively decoupled spectrum from one that is nondecoupled.

* The selective population inversion (SPI) (48, 49, 50, 51) method applies a 180° pulse to a selected transition of nucleus X with the decoupling channel; nucleus A is then observed with a 90° pulse. The 180° pulse must be generated by a radio frequency field B_2 and $\gamma_X B_2/2\pi << J_{AX}$ (soft pulse). The same pulse sequence is valid for homonuclear SPI (52) apart from the 90° observing pulse, which is replaced by a $\pi/6$ pulse (Ernst condition, 28).

3.8.4 Calibration of the Decoupler Radio Frequency Field

Some of the decoupling experiments described need a precise decoupler power setting. The decoupler power can be calibrated by measuring a residual heterocoupling (e.g., ^{13}C-{H} in TMS) J_r with respect to the offset $\Delta\nu$ (Hz) of the decoupler carrier frequency from the X multiplet under decoupling.

Then

$$B_2 = \frac{J_{real} \; 2\pi\Delta\nu}{\gamma_X \cdot J_r}$$

This method is difficult to apply in the homonuclear case because of the Bloch-Siegert effect (Section 3.8.2). The decoupling channel can then be used as an observation channel. A nulled signal (180° pulse) for a chosen X resonance will identify B_2 as

$$B_2 = \frac{\pi}{\gamma_X \cdot \tau_{180}}$$

In this experiment the decoupler frequency offset must be set on the X resonance τ_{180} represents the time during which the decoupler is on, and the FID is acquired, of course, with no excitation pulse from the normal emitter.

3.9 BROAD LINES, LOW FREQUENCIES

3.9.1 Detection

A broad line with an associated short T_2 develops from a fast decay following the excitation pulse. Because sometimes the beginning of this FID is not acquired for electronic reasons (cf. Section 3.7.1), the information is lost. A simple way to get a flat spectrum baseline with as much information as possible is to use an echo sequence (53; B11, p.23), provided that the delay τ between the $90°$ and $180°$ train is $\tau < T_2$.

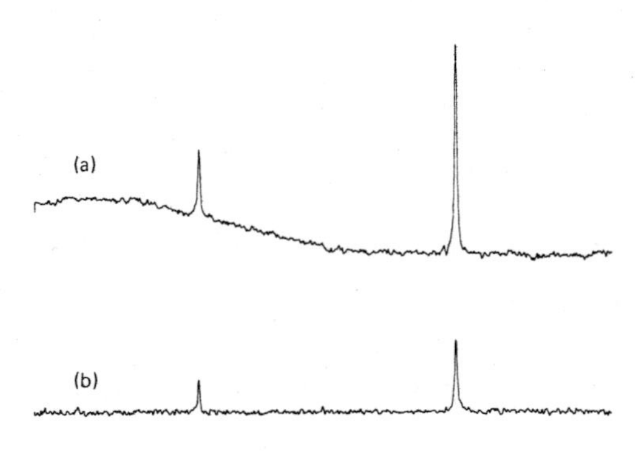

Figure 23. Acoustic ringing elimination via a ($90°$-τ-$180°$-τ-acquire) sequence. ^{17}O spectrum of a (H_2O + acetone) mixture: (a) normal spectrum; (b) echo sequence with τ = 50 ms.

When a residual anisotropic interaction (dipolar for spins 1/2, quadrupolar for spins > 1/2) does exist this echo sequence is modified as $90°_x$, τ, $90°_y$, τ-acquire (54, 55).

3.9.2 Line Position

If no fine structure exists, the dispersion mode is preferred to define the line chemical shift more precisely.

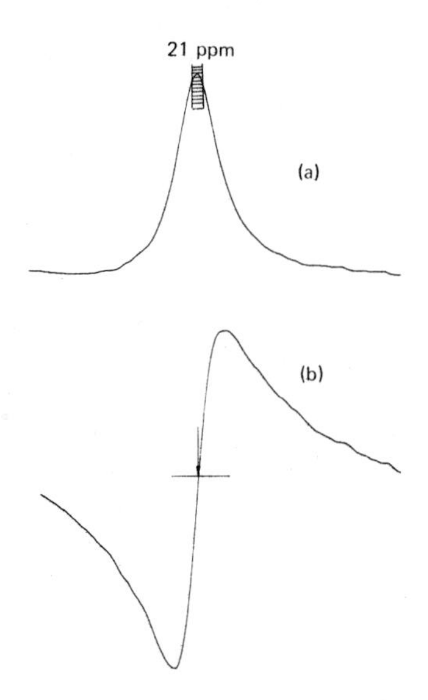

Figure 24. ^{187}Re spectrum of ReO_4^- ($\Delta\nu$ = 4000 Hz). The absorption mode spectrum (a) gives a ± 11ppm accuracy on line position. Dispersion mode (b) gives much better precision.

3.10 ELIMINATION OF UNWANTED LINES

When information that requires discrimination between weak satellite lines centered around a main resonance is sought or when some unwanted resonances fall within the explored chemical shift region, the following methods can be tried.

1. Unwanted peak saturation (45, 56) or presaturation (57) by the decoupler channel or cross saturation by exchanging sites (58).
2. Selective pulse sequence (63, 64); the selective 90° or 180° pulse on the unwanted peak is followed by a nonselective 90° pulse (this sequence has already proved to be useful in ^{31}P-NMR).
3. Discrimination of several lines by their relaxation times T_1 (59, 60, 61) with a (180°-τ-90°-acquire)

sequence and adjusting τ to null the desired line.

4. Discrimination of several lines by their relaxation times T_2 (62, 66), with a (90°-τ-180°-τ-acquire) sequence and ajusting τ for nulling the desired broad line (Figure 25) or a (90°-τ-180°-τ-90°-acquire) and choosing τ to null the desired sharp lines (71).

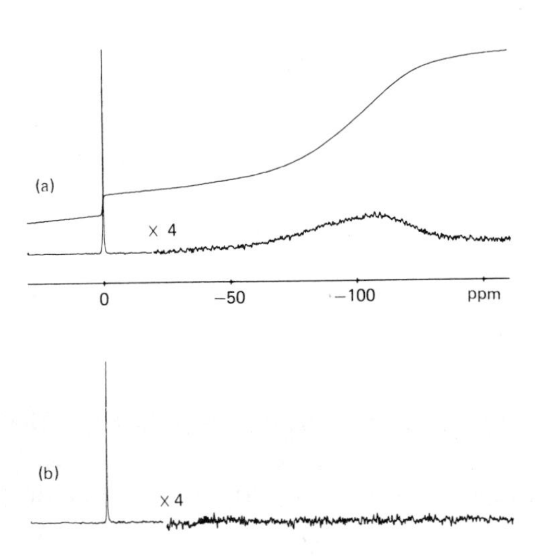

Figure 25. Elimination of ^{29}Si signal background from (insert + tube) via a (90°-τ-180°-τ-acquire sequence). Sample: TMS: (a) normal acquisition mode; (b) echo sequence (τ = 50 ms).

5. Decoupling plus difference spectrum techniques (68, 69, 70), which allows the observation of weak satellite lines on each side of an intense line.

3.11 SPECIAL PULSE SEQUENCES

These pulse sequences are actually the most promissing experiments because they can be used with any nucleus.

3.11.1 Selective Impulsion Train: DANTE sequence (64, 72)

The desired (90° or 180°) pulse is obtained as the sum of n micropulses of duration $\tau_n = \tau_{90°}/n$ or $\tau_{180°}/n$ with repetition time t_r s. The excited frequency domain is then made of discrete frequency bands $\nu_i = \nu_0 \pm i/t_r$ with a narrow half-width $\Delta\nu_{1/2} \sim 1.2/(n.t_r)$ (Figure 26).

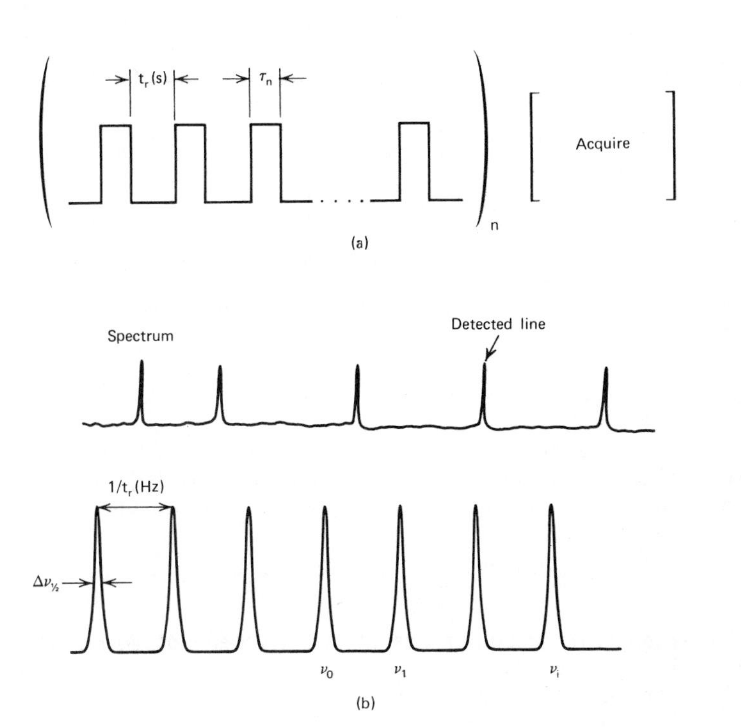

Figure 26. Simulation of a selective pulse train: (a) time domain; (b) frequency domain.

Several applications can be foreseen for such a sequence:

Peak suppression (cf. Section 3.10.2)

Selective excitation of a given resonance in a complex spectrum and acquisition of its FID under chosen decoupling conditions (Figure 27).

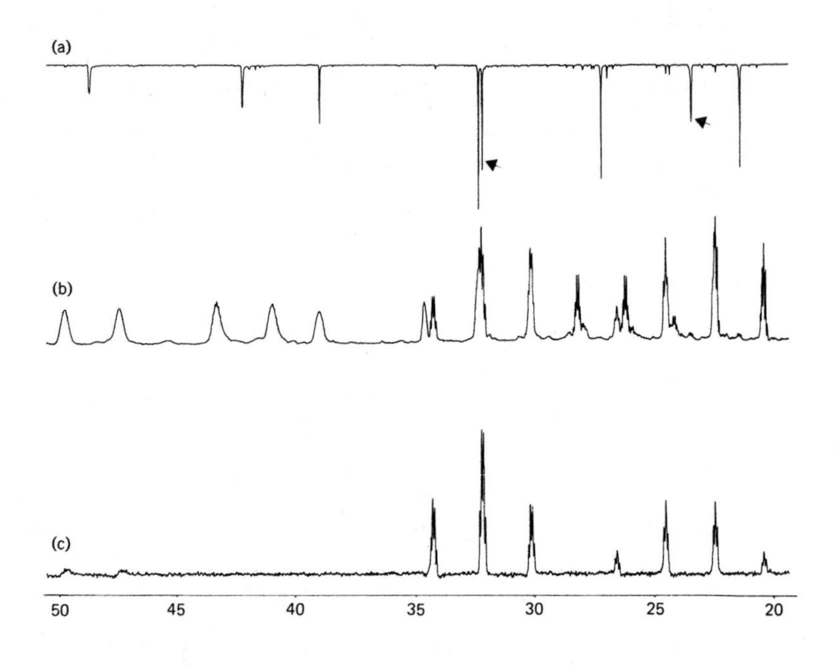

Figure 27. ^{13}C selectively picked absorptions (SEPIA) sequence. Sample α pinene: (a) normal decoupled spectrum; (b) normal proton coupled spectrum; (c) SEPIA proton coupled spectrum. (Bruker Report 1/1980).

Exchange-rate measurements by the Forsen-Hoffman (73) or Dahlquist (74) methods.

Relaxation time measurements.

3.11.2 Spectrum Acquisition Speeding

The following methods can be tried when observing nuclei that suffer from long T_1 values which need long intervals between acquisition pulses.

Driven Equilibrium Fourier Transform (DEFT) (75, 76): Spin Echo Fourier Transform (SEFT) (78). These two methods are based on echo-type refocusing pulse sequences. The theoretical gain in sensitivity was originally claimed to be as high as 20 with respect to conventionnal acquisition, but a critical experimental screening (77) clearly showed that this enhancement factor was only around 2 (107).

Homo Spoil (80). In this method any magnetization left after the end of the acquisition period in the plane perpendicular to B_0 is suppressed by "spoiling" briefly one of the homogeneity gradients.

3.11.3 Pulse Shaping

Some experiments are sensitive to pulse imperfections. Recent paper (81) reviewed the problem completely and furnished elegant solutions to the following:

B_1 inhomogeneities.

Offset- and phase-dependency compensation in multiline spectra.

Precise obtention of a 180° tilt angle.

3.12 SENSITIVITY ENHANCEMENT

Many low γ nuclei suffer from a lack of sensitivity. Some recently proposed pulse sequences based on a detectable J_{A-X} scalar coupling can be envisaged to overcome such a problem, apart from the standard decoupling procedures that provide a sensitivity gain in coalescence of any multiplet into a single peak with or without an intra or intermolecular NOE build up (82; Section 1.5.4).

3.12.1 SPI or SPT Methods (see Section 3.8.1 and 3)

The corresponding theoretical gain to be expected (39) amounts to

$$\frac{\gamma_A \pm \gamma_X}{\gamma_A}$$

where A is the observed and X is the irradiated nucleus. When $T_{1X} < T_{1A}$ (e.g., A = ^{15}N, X = ^{1}H) the FID acquisition is speeded up as the optimum interpulse delay is set roughly equal to 1.25 T_{1X} (39, 42, 83). The promising INEPT sequence (84) has also been proposed.

$$[90^\circ_x(X) - \tau - 180^\circ_x(X) - 180^\circ_x(A) - \tau - 90^\circ_y(X) - 90^\circ_x(A) - \text{acquire}]_n$$

where $\tau = 0.25/J_{AX}$.

3.12.2 J Cross-Polarization Method (JCP)(85, 86)

This method, although not yet well documented on experimental details, has been proposed, to parallel the well-known cross-polarization experiments in solid-state NMR. A selective 90° pulse is applied to the X nucleus resonance and followed by a nonselective pulse on the A nucleus, with the Hartman-Hahn (86) condition fulfilled; that is

$$\gamma_X B_2^{(X)} = \gamma_A B_1^{(A)}$$

where $B_2^{(X)}$ and $B_1^{(A)}$ are the strenghts of the decoupling and observing radio frequency fields respectively.

3.13 QUANTITATIVE MEASUREMENTS (B14, B18)

Any NMR experiments can be foreseen as quantitative, provided that some factors are considered (87,88) or weeded out (89, 90) with paramagnetic reagents. Optimal conditions for these measurements may be found in Ref. 91. A pulse sequence may also be used (82, 88; B20, p.231).

3.14 SAMPLE TEMPERATURE MEASUREMENT

Apart from the study of the J or δ dependency on temperature, this parameter is of prime importance when dealing with exchange or conformationnal processes: for example, the determination of the relevant thermodynamical constants

($\Delta_G^{\neq}$, $\Delta_S^{\neq}$) of these dynamic pathways.

In any case, ±0.5°C appears to be the lowest limit that can be expected as temperature gradient in a 10-mm tube (92). On the other hand, pulse trains or decoupling experiments do warm up the sample, especially with ionic solutions; therefore the sample should stand in the probe at least 10 min under the measuring conditions to allow for sample temperature equilibration before temperature measurements or data acquisition are started.

In addition to measuring the temperature with the variable temperature unit installed on the spectrometer (accuracy ±2°C), the following methods can be used:

3.14.1 Direct Measurement

By introducing a calibrated temperature sensor directly into the tube (e.g., thermometer, thermocouple, or thermistor); this method precludes tube spinning. Moreover, a thermocouple or a thermistor, when placed in a magnetic field, can undergo severe secondary effects that lead to serious errors in temperature measurement (94).

3.14.2 Indirect Measurements

For indirect measurements the calibrated temperature chemical shift dependence of a given compound can be used. This method implies sample tube exchange and temperature equilibration. The universal method that involves ethylene glycol or methanol proton chemical shifts appears to be the simplest and most accurate indirect experiment, provided that a proton spectrum can be recorded by the decoupling coils of the multinuclear probe. Some calibrated samples have also been studied for ^{19}F (95, 96) or ^{13}C (97, 98, 99). Levy et al. (100) devised a sensitive NMR temperature sensor in the ^{13}C frequency range: namely, the ^{59}Co resonances (1.504 ppm/°C for $Co(CN)_6$, 3.153 ppm/°C for $Co(acac)_3$). These cobalt-chemical shifts were perhaps too temperature-sensitive but other standards have been successful (9).

An original method for obtaining a probe temperature calibration curve which uses the sharp melting point of known samples has been published. The solid-liquid phase transition is detected visually (103) or by NMR (101, 102).

CHAPTER 4

DYNAMIC MEASUREMENTS

Heteronuclei NMR will in the near future certainly prove to be useful in extracting dynamic parameters otherwise difficult to get from proton or carbon 13 studies. As an example, because of their dominant quadrupolar relaxation, aluminium, deuterium, oxygen 17, and nitrogen 14 are ideal candidates for T_1 measurement, hence correlation time determination; a low γ spin 1/2 nucleus can be a very sensitive probe in a proton environment by selective proton heterodecoupling and NOE measurements, and it can be safely predicted that homonuclear NOE determination on nuclei other than proton by enrichment techniques will soon appear. This chapter therefore presents the techniques, methods, and drawbacks of these dynamic experiments.

4.1 NUCLEAR OVERHAUSER EFFECT (A NUCLEUS OBSERVED, X NUCLEUS IRRADIATED)

The NOE was introduced in Section 1.5.4 and the relevant sample preparation conditions were developed in Section 3.1.2.

4.1.1 Homonuclear NOE measurement

Two FID are acquired, one with the decoupler on, then gated off during acquisition, the other with the decoupler continuously off. The same recording conditions (number of scans, waiting time, memory size, pulse angle, etc.) are obviously required to compare the two transformed FID. The decoupling frequency can also be shifted a few hundred Hertz away from the on resonance conditions when the "nonirradiated" spectrum is recorded to keep the computer duty cycle unchanged between the two experiments. Recently developed computers also allow to perform in core NOE difference spectra by alternatively substracting the irradiated FID from the nonirradiated (108). This method is useful in complex spectra because only the lines experiencing NOE enhancement, as well as the irradiated are detected.

4.1.2 Heteronuclear NOE

A spectrum is recorded under continuous X nucleus broadband heterodecoupling; a second is then acquired with the decoupler gated on just before the acquisition. No NOE develops during this second irradiation (109). To avoid any sample temperature change it is strongly recommended that the decoupler frequency offset be shifted a few kilohertz instead of gating the decoupler off-on (110).

4.1.3 Waiting time

All NOE measurements imply that the A spin system is in equilibrium. This condition requires a pause before each acquiring pulse at least 10 times longer than longest T_1 of the irradiated or observed nuclei (109, 111, 112). On the other hand, the NOE is field and correlation time dependent (see Section 1.5.4).

4.1.4 Precision

Regardless of the experimental conditions, the precision expected in a NOE determination will never exceed a few percent (109, 111, 112). The final signal-over-noise is also of prime importance (115). As an example, a signal-over-noise ratio of 140/1 is required to obtain a 10 % η factor with a 10 % error.

4.1.5 Precision Improvement: Repetitive Frequency Shift Method (113, 114)

In this RFS method the irradiated and nonirradiated FID are recorded repetitively with an alternative shift in the carrier frequency to obtain one unique FID that, when transformed, will provide simultaneously the desired normal and NOE enhanced lines and allow for better integration.

4.1.6 NOE Quenching

The NOE is sometimes annoying, especially in quantitative measurements. It can be eliminated

* with special pulse sequences (B20, p.231);
* by using paramagnetic relaxagents (89, 90, 116).

4.1.7 Truncated or Driven NOE (TOE, DNOE)

The rate of NOE growth is dependent upon the distance between observed and irradiated nuclei. In some cases it is interesting to study this time dependence of NOE establishment, which allows the molecular dipolar environment of the A nucleus to be described by the X nucleus.

Two methods have been developed for this purpose;

* Truncated NOE (117). The following pulse train is applied to the spin system:

$$\left[(180°_{(X)}, \tau_i, AQT_{(A)}, \tau_2) - (180°_{(off.res.\ X)}, \tau_i, AQT_{(A)}, \tau_2)\right]_n$$

where τ_i is a variable delay and τ_2 is a waiting interpulse time ($10T_1$).

* Radiofrequency driven NOE (118). The following sequence is applied:

$$\left[(X\ irrad.\ during\ time\ t_i,\ AQT_{(A)}, t_2) - (off\ res.\ X\ irrad.\ during\ time\ t_i,\ AQT_{(A)},\ t_2)\right]_n;$$

t_i and t_2 play the same roles as τ_i and τ_2 in a TOE experiment. In both methods a plot of η_i versus τ_i or t_i is then drawn for each A lines.

4.1.5 NOE and T_1

The NOE growth associated with the X nucleus irradiation period is exponentially related to the total relaxation time T_1 of the A nuclei (109, 110), but its final value depends only on the A dipolar relaxation time T_{1dd} (109). This situation permits a simultaneous determination of η_{A-X} and T_{1A} (119, 120, 121). A NOE measurement by the selective irradiation of one transition of a complex spin system can yield a wealth of informations on the different relaxation mechanisms of this spin system (122). Last but no least this

A-species NOE dependence on X nucleus irradiation is one of the few ways of measuring T_{1A} under X decoupled conditions when for some reason (temperature, viscosity, etc.) η_{A-X} is exactly equal to -1(123).

4.2 T_1 MEASUREMENTS

The available methods of T_1 measurement are delineated only with respect to pulse sequence and operating conditions.

4.2.1 Inversion Recovery Fourier Transform (IRFT) (124, 125)

This method uses the following sequence:

$$(180°\text{-}\tau\text{-}90°\text{-acquire-t})_n$$

where τ is a variable delay and t is a recovery time chosen as (acquisition time + t) $\geqslant 5T_1$. The measured line intensity is expressed as

$$I_\tau = I_\infty \left[1\text{-}2 \exp\left(-\ \tau/T_1\right)\right]$$

where I_∞ corresponds to $\tau = \infty$ (spin system in equilibrium), a condition practically met with $\tau_\infty = 5T_1$ (Figure 28).

This method seems to be the most reliable, provided that some experimental precautions, carefully reviewed (112), are kept in mind. If an estimate of the T_1 value can be made a 10 τ experiment will span a range between $0.3T_1$ and $2T_1$.

Modifications and Improvements

- The phase of the observing 90° pulse can be inverted alternatively (126) to avoid echo formation at small τ values ($T_2^* \sim T_1$).
- Freeman-Hill modification (125): the acquired FID corresponds to $(I_\infty - I_\tau)$ in the sequence.

$$\left[(90°, \text{acquire}, t) - (180°, \tau, \text{acquire}, t)\right]_n.$$

Instrumental time can become prohibitive for long T_1.

•Fast inversion recovery (FIRFT) is a method in which the long waiting time t between each sequence (127, 128) is replaced by a shorter t', thus speeding up the measurement. The line intensity obeys the following equation:

$$I_\infty - I_\tau = I_\infty \left[2\text{-exp}\left(-\frac{t'w}{T_1}\right) \right] \cdot \exp(-\tau/T_1)$$

where t'w = (t' + AQT) and t'w $\geqslant 2T_1$. A dynamic steady state which requires the elimination of the first 10 scans must be reached before each acquisition. On the other hand, this method is sensitive to pulse value missetting and carrier offset position. A three-parameter, direct exponential fit is strongly recommended (129) to extract T_1.

•Modified fast inversion recovery (130). In this case the (t, τ) values are set as

$\tau + t = \Delta$ = constant, with $t \geqslant 3T_2^*$ and $\Delta \sim 3T_1$ which is useful for small T_1 values.

•Inversion recovery spin echo (131) permits T_1 measurements on sharp (long T_1 lines) among broad (short T_2) ones following the sequence

$$(180°, \tau_1, 90°, \tau_2, 180°, \tau_2, \text{acquire}, t)_n$$

where τ_1 is a variable delay, τ_2 is chosen to exceed the (short) T_2 values of the broad lines which will be worn off during the (90°-τ_2-180°-τ_2) echo sequence.

4.2.2 Saturation Recovery Fourier Transform (SRFT)

Method (132, 133). The 180° pulse of the IRFT sequence is replaced by a 90° pulse (or a 90° pulse train); that is,

$$\left[(90°)_i - \tau - 90° - \text{acquire}\right]n,$$

which corresponds to the following line intensiy recovery law (Figure 29):

$$I_\tau = I_\infty \left[1 - \exp(-\tau/T_1)\right]$$

An homospoil pulse can be inserted between the 90° pulse

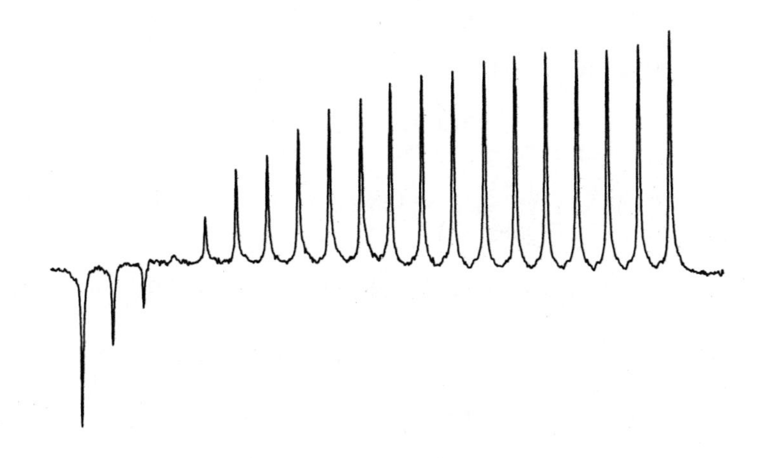

Figure 28. 20.261 MHz ^{99}Tc T_1 measurement by the IRFT-RFS method. Sample TcO_4^-; 8 scans, T_1 = 140 ms.

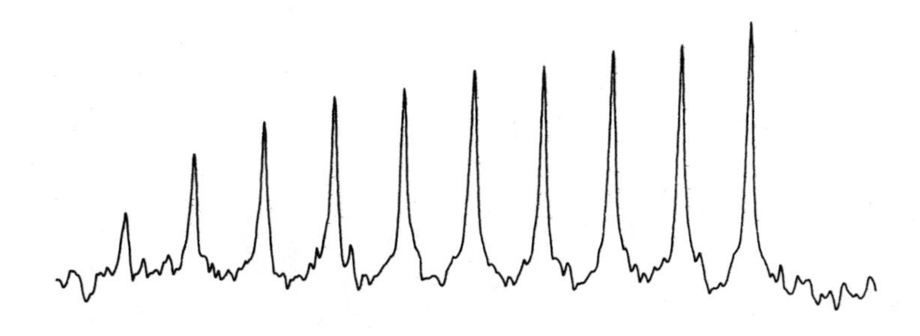

Figure 29. 33.55 MHz ^{119}Sn T_1 measurement with the SRFT-RFS method. Sample: undegased $SnMe_4$; 20 scans: T_1 = 580 ms.

train $(90°)_i$ and the variable delay τ and again after FID acquisition to avoid echo build up or an "aperiodic pulse sequence" (134) can be used.

This method is sentitive to pulse angle value missetting.

Modification. The 90° saturating pulse train can be replaced by any spin system saturation method (132).

4.2.3 Progressive Saturation Fourier Transform (PSFT) (135) (Figure 30)

This method is based on a $(90°\text{-acquire-}\tau)_n$ sequence. The first 5 to 10 scans are deleted to reach a spin-system steady state. The exponential line intensity recovery is then expressed as

$$I_\tau = I_\infty \left[1 - \exp \left(- \frac{AQT + \tau}{T_1} \right) \right]$$

This method obviously excludes measurements of samples for which T_1 are of the order of magnitude or below of the chosen acquisition time.

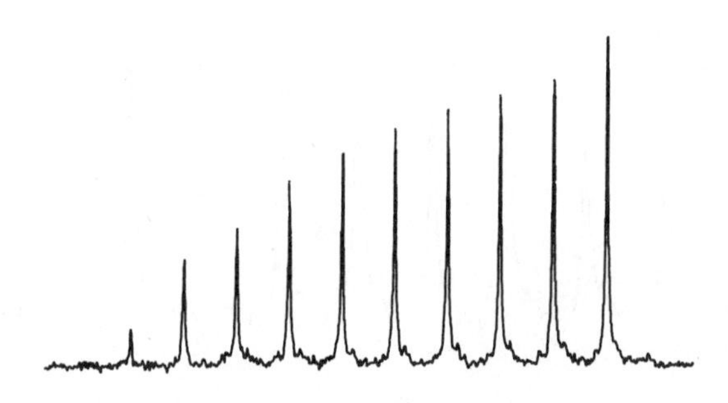

Figure 30. 22.63 MHz ^{13}C T_1 measurement with the PSFT-RFS method. Sample $TeMe_2$. Dummy scans: 5, T_1 = 18.5 s.

4.2.4 Variable Nutation Angle Fourier Transform (VNAFT)

Instead of using a *variable delay* after the 90° preparative pulse or pulse train, these VNAFT methods associate a *variable pulse angle* with a *fixed delay*. They are, nevertheless, critical to handle and do require a precise pulse value calibration and a complete mathematical treatment of the measured intensities. Only relevant methods taken from the literature are cited.

PSFT-Derived Experiment (136, 137, 138). This experiment can be used for time-dependent spin systems (e.g., CIDNP experiments).

IRFT-Derived Experiment (139).

4.2.5 Single-Scan Fourier Transform (SSFT) (140)

The spin system magnetization is inverted by a 180° pulse, then sampled during its return to equilibrium with Ni successive (θ° pulse-acquire) sequences. Each acquisition is stored on a disk and the process is repeated as many times as necessary to improve the signal-over-noise ratio of the acquired spectra.

The sequence is the following

$$\left[180°,\ \text{spoil},\ (\theta°,\ \text{acquire},\ \text{spoil},\ t)_{Ni},\ t'\right]_n$$

Where t is the overall disk access and transfer time and t' is the waiting time. This method is useful only for long T_1 values (> 5 s) or for T_1 measurements on reactive species. A variant is the repetitive frequency shift method (113) in which no mass storage device is needed.

4.3 EXPERIMENTAL PRECAUTIONS AND ERRORS IN T_1 MEASUREMENTS

4.3.1 Data Handling

The T_1 value is obtained by fitting the measured line intensities and the corresponding variable parameter (τ_i or $\theta°_i$) to the theoretical intensity recovery law which corresponds to the used method. The semilog plot is, in fact, strongly warned against (141, 142, 153, 144); a least-square fit is preferred, but surpassed by a multiparameter direct exponential fit (141, 145).

4.3.2 Experimental Errors

Sample Preparation (see Section 3.1.2)

Pulse Angle Value Adjustment (see Section 3.6.4). A precise 90°or 180°pulse value can be achieved in a "composite pulse" sequence (81).

Carrier Offset. The offset condition, with respect to the carrier frequency, of any T_1 experiment will lead to systematic errors (142, 144) which can be counterbalanced by a few tricks (81b). The IRFT method is the least sensitive to this offset effect.

Exciting Radio Frequency Field Inhomogeneities. This effect is more severe with single-coil probes than with cross-coil probes. The inhomogeneities require careful sample preparation (cf. Section 3.1.2). A multiparameter fit (141, 142, 146) can also compensate in part for these side effects. A good test for the RF field homogeneity can be set up by measuring a line intensity I after 90° and 270° pulses (141, B13, p.131). The ideal result is $I_{90°} = -I_{270°}$.

Acquisition. A T_1 measurement requires a good setting from the spectrometer receiver gain to realize full use of the available dynamic range. The phase of the transformed spectrum should also be carefully adjusted for correct intensity determination.

4.3.3 Choice of the Measuring Method

Each method possesses its own domain of selectivity. Nevertheless, the IRFT appears to be the least sensitive to systematic errors (144, 145) and is available with every standard commercial software package. A sensible precaution before starting any T_1 measurement, whatever the method chosen, is to run a standard sample that will permit all relevant parameters to be set up precisely.

4.4 T_2 MEASUREMENTS

T_2 measurements are much more demanding than T_1 experiments with respect to spectrometer performances and sample characteristics (B13, p.131) and are based on echo formation.

4.4.1 Carr-Purcell Method

The following sequence is established (79):

$$\left[90°,\ \tau,\ 180°,\ \tau,\ \text{acquire},\ t,\ 180°,\ \tau,\ \text{acquire},\ t,\ldots,\ t'\right]_n$$

where (AQT + t) = τ; t' is the spin system recovery time. When $T_1 > T_2$ t'is then T_1 dependent. This sequence is difficult to achieve with a conventional high-resolution Fourier spectrometer.

4.4.2 Meiboom-Gill Modification of The Carr-Purcell Sequence (CPMG Sequence)

This modification avoids the cumulative pulse misadjustment effects over the entire sequence (147) by shifting the 180° pulse phase of $\pi/2$ with respect to the initial 90° pulse; it also compensates in part for any exciting RF field inhomogeneities or offset effects.

4.4.3 Fourier Transform Method

Because it is generally impossible to acquire several FID after one single 90° pulse, especially for $T_2 < 5$ s, Fourier transform T_2 measurements are handled by a modification of the Meiboom-Gill method in which p.180° pulses, equally spaced by 2τ s, are generated after the initial (90°-τ-180°) burst. At time τ following the pth 180° pulse the second half of the echo is acquired and transformed. Varying p permits sampling of the magnetization decay hence measurement of T_2:

$$I_{\tau_p} = I_o \exp(-\tau_p/T_2)$$

where I_o is the signal intensity at equilibrium and τ_p represents the time interval between the initial 90° pulse and the half-echo acquisition start $\tau_p = \tau.(p + 2)$. The best results are obtained with τ values around 100 msec. Of course, the desired half-echo is acquired with a null pulse angle and no delay occurs between this triggering blank pulse and data acquisition. Care should be taken to *avoid any kind of alternated phase sequences during acquisition* because they are devised primarily for echo suppression!

Finally, any phase glitch in the transformed spectrum can be eliminated by using the power spectrum mode (cf. Section 2.6.4.) Figure 31 exemplifies a T_2 measurement.

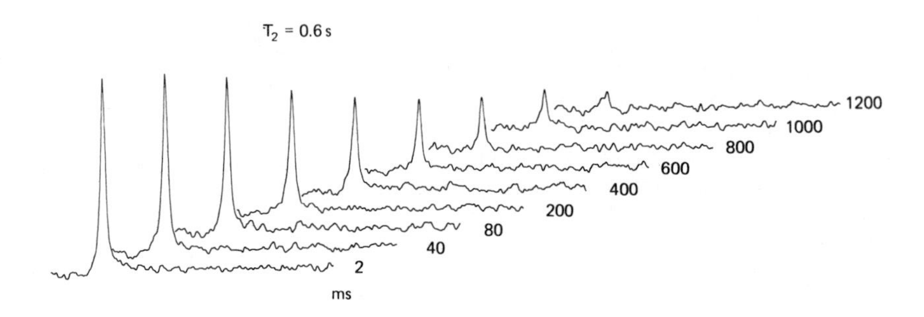

Figure 31. 16.3 MHz ^{95}Mo T_2 measurement via a CPMG pulse sequence. Sample: MoO_4^{2-}, M in D_2O.

4.4.4 Experimental Errors and Limits of the Method (One-Line Spectrum)

The precautions that hold for T_1 measurements in regard to sample preparation are valid when T_2 experiments are run. A detailed study will be found in Ref. B31, Vol.5, p.117, and Refs. 148, 149, 150, and 151. The following should be remembered:

- External lock must be rejected for T_2 measurements (13).
- Spinning the sample leads to a modulation in the echo decay (152, 153). If rotation of the sample is unavoidable because of resolution requirements, short τ values that are not a multiple of the spinning rate are mandatory.
- When heterodecoupling is used the value of the decoupling field B_2 will be set (154) to avoid the Hahn condition (86); namely, $(\gamma_A^{obs} \cdot B_1) = (\gamma_X^{dec} \cdot B_2)$. In broad-band hetero-

decoupling this condition can be met. To eliminate the undesirable resulting effect noise decoupling can be switched off during echo formation and reestablished just before acquisition and during waiting time to benefit from any potential NOE effect (B13, p.131).

- T_2 value depends directly on the dynamics of the spin system under study (chemical exchange, etc.). Its measurement provides access to dynamic parameters.
- Last but not least, T_2 value can be determined by measuring the resonance half-width, provided that the relaxation of the isotope under study is dominated by a quadrupolar mechanism under extreme narrowing conditions; then,

$$T_2 = \frac{1}{\pi \Delta \nu_{1/2}}$$

4.4.5 Multiline Spectra

The Meiboom-Gill method is equally valid for multiline spectra when no homonuclear J coupling occurs. If couplings exist, they introduce an echo J modulation (153), which can otherwise be used for precise J measurements. This undesirable modulation effect can be reduced

- by using a short τ value such as $| 2\pi J.\tau | << 1$, where J corresponds to the highest homonuclear J value;
- by exciting selectively and observing one line at a time (154). This is achieved by choosing a selective digital filter (154) or by performing the selective pulse sequences presented in Section 3.11.1 (64, 72).

4.5 SIMULTANEOUS MEASUREMENT OF T_1 AND T_2

4.5.1 Kronenbitter and Schwenk Method (105, 106, 155)

This method measures T_1/T_2 and $1/2\ (T_1+T_2)$ simultaneously. Schwenk et al. have used this procedure for low γ nuclei with long T_1 and T_2 values that give rise to only one resonance. It needs a rather modified spectrometer.

4.5.2 Eisenstadt Method (156)

This method can be used for only single-line spectra and is derived from the triplet sequence (157). It needs a precise pulse adjustment and good magnetic field homogeneity.

4.6 SPIN LOCKING $T_{1\rho}$ MEASUREMENTS

In the liquid state in general $T_{1\rho} = T_2$. This $T_{1\rho}$ measurement requires less spectrometer strain with respect to pulse setting but does necessitate special probes when long (> 2 sec) $T_{1\rho}$ are measured, due to the high power needed to align the spin system along the rotating field B_1.

4.6.1 Freeman-Hill Method (158, 159)

The operating sequence is the following :

$$\left[90^\circ_x, (B_{1y})\tau_i, \text{acquire}, t\right]_n$$

where B_{1y} represents a RF field in quadrature with the 90° pulse, which is applied during a time τ_i and t is the equilibrium recovery time. If broad band heterodecoupling is used, the precautions needed for T_2 measurements hold (cf. Section 4.4.4).

$T_{1\rho}$ is obtained as follows :

$$I_{\tau i} = Io \exp(-\tau_i/T_{1\rho})$$

This method is not sensitive to pulse maladjustment and does not suffer from offset effect (160, 161). It has been used to study exchange rates between 10^3 and 10^5 sec^{-1} (160).

4.6.2 $T_{1\rho}$ Off Resonance Modification (162)

This method is demanding of probe design because the locking field B_{1y} is applied off resonance during a fixed τ time (at least $5T_1$), and followed by an homospoil pulse and a 90° on-resonance acquisition. It permits the measurement of long $T_{1\rho}$ values.

4.6.3 Irradiation by Narrow Frequency Envelopes by Repeat Nutation Orbiting (INFERNO) Sequence (72)

This promising selective spin-locking experiment, run on a multiline spectrum, can be performed on any commercial high resolution FT spectrometer. The corresponding sequence is the following :

$$\left[(\alpha_x, \tau)\, m,\ (\beta_y, \tau) m',\ \text{acquire},\ t\right]_n$$

where $m.\, \alpha_x = 90°$. Generally m = 100 and the time interval τ is chosen to excite one line selectively (cf. Section 3.11.1). The second part of the pulse sequence represents the spin-lock pulse with m' in the order of several thousands. Varying m' corresponds to the variable τ values in a conventionnal spin-locked experiment.

PART TWO

ISOTOPE NMR FINGERPRINTS

The second part of this book will provide the NMR spectroscopist with all the information required to start experiments on a given nucleus. For each magnetically active, diamagnetic isotope (except ^{1}H, ^{19}F, ^{13}C, ^{40}K) the following is supplied:

1. A test sample (compound, solvent, concentration) and the corresponding spectrum with its relevant acquisition and processing parameters.
2. The line frequency in a field corresponding to exactly 100 MHz for the protons of TMS (Ξ value).
3. A chemical shift scale for the nucleus under study for a quick estimate of where an unknown sample should resonate and characteristic T_1 values (an important parameter for deciding on pulse repetition rate and pulse angle value).

All spectra have been run with 10-mm tubes unless otherwise stated (* = 15-mm tube). In a few cases no resonances could be recorded, due certainly to very broad lines (^{177}Hf, ^{179}Hf, ^{191}Ir, ^{193}Ir, ^{197}Au, ^{201}Hg). Nevertheless as much information as possible has been provided for a beginning on a study of these isotopes.

In the Lanthanides series Pr (Z = 59), Nd (Z = 60), Pm (Z = 61), Sm (Z = 62), Eu (Z = 63), Gd (Z = 64), Tb (Z = 65), Dy (Z = 66), Ho (Z = 67), Er (Z = 68), Tm (Z = 69) are found in an almost paramagnetic state and it is thus unlikely that high resolution NMR work can be used for these nuclides. On the other hand, ^{70}Yb and perhaps ^{71}Lu will certainly prove useful. For ^{70}Yb only the +II oxydation

state yields diamagnetic (although unstable) compounds; for ^{71}Lu the +III oxydation state produces a vast family of diamagnetic compounds.
The actinides [Ac (Z = 89), Th (Z = 90), Pa (Z = 91), U (Z = 92), Np (Z = 93), Pu (Z = 94), Am (Z = 95), Cm (Z = 96), Bk (Z = 97), Cf (Z = 98), Es (Z = 99)] have been deliberately withdrawn from the tables. In fact, only ^{92}U should be amenable to NMR experiments, although its large quadrupole moment (3.18 Barn) associated with a low resonance frequency ($\Xi \sim 1.76$ MHz) will render them rather acrobatic.

The parameters are expressed in

Concentration: mole/liter
B_o: Tesla
- 2.114 T corresponds to 90 MHz for ^1H
- 5.875 T corresponds to 250 MHz for ^1H

Temperature: degrees Celsius

Gyromagnetic ratio: $\gamma/10^7$ rad. T^{-1} . s^{-1}

Quadrupole moment: $Q/10^{-28}$ m^2. A (-) sign indicates an unknown quadrupole moment.
The repetition rate corresponds to: (acquisition time + inter pulse delay).
All decoupled spectra are proton broadband decoupled; any potential negative NOE has not been considered in spectrum phasing.
$\Delta\nu$ is the line width at half-height.

HYDROGEN Z 1

Sample	: H_2O (^3H, ^2H)	Lock	:	^2H
Solvent	: D_2O	Bo	:	2.348 (^3H)
Concentration	: dif. for ^2H & ^3H	Temperature	:	35°C

^3H

Spin	: 1/2
Nat abund (%)	: -
Receptivity / ^{13}C	: -
Gyromagnetic ratio	: 28.533
Quad. moment	: -

Resonance Freq. (MHz) referred to ^1H TMS resonance at 100MHz

106.663

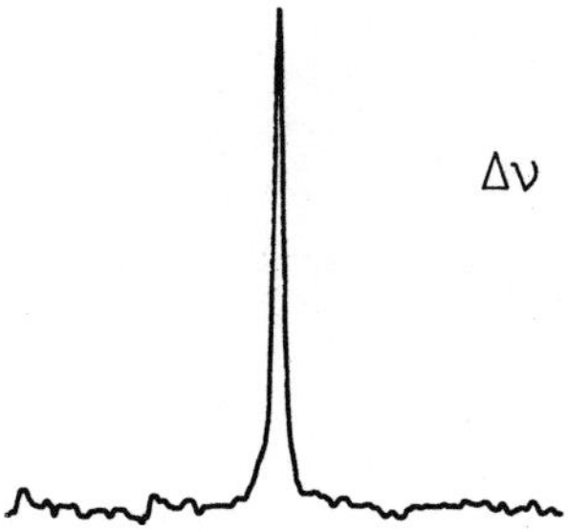

Δν : 0.9 Hz

Spectral width (Hz)	: 1430
Number of data points	: 16 K
Pulse	: 4 μs/60°
Number of scans	: 10
Repetition rate (s)	: 1.4
^1H decoupled	: yes
Exp. filter (Hz)	: 0

^2H

Spin	: 1
Nat. abund (%)	: 0.015
Receptivity / ^{13}C	: 8.2 10^{-3}
Gyromagnetic ratio	: 4.1064
Quad. moment	: 2.73 10^{-3}

Resonance Freq. (MHz) referred to ^1H TMS resonance at 100MHz

15.350_7

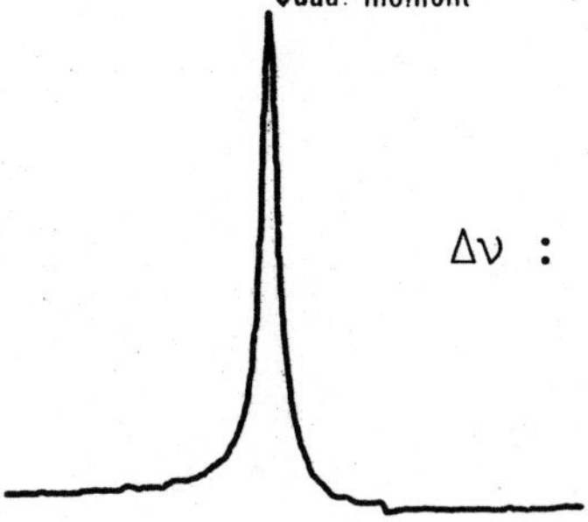

Δν : 1.3 Hz

Spectral width (Hz)	: 600
Number of data points	: 4 K
Pulse	: 5 μs/ 10°
Number of scans	: 1
Repetition rate (s)	: -
^1H decoupled	: -
Exp. filter (Hz)	: 0

CHEMICAL SHIFTS : reference TMS

For diamagnetic compounds : $10 < \delta < 0$
For paramagnetic compounds : $- 500 < \delta < 500$.
cf : NMR of Paramagnetic molecules
Academic Press (1973)

ppm

COUPLING CONSTANTS : typical values (Hz)

	^1H	^{19}F	^{31}P	^{13}C	Homo
$\lvert ^1J \rvert$ (^1H)	- to -	530 (HF)	200 to 1100	96 to 320	278 (H_2)

RELAXATION :

T_1 typical values (s) : ^2H : 0.1 to 10

^3H : 1 to 10

LITTERATURE :

B. 12 p. 107

I.C.P. Smith et al. : Prog. NMR Spectry. <u>11</u>, 211 (1977)

REMARKS :

LITHIUM Z 3

Sample	: LiCl	Lock	: ^2H
Solvent	: D_2O	Bo	: 5.875
Concentration	: 1 M	Temperature	: 27°

^6Li

Spin	: 1
Nat abund (%)	: 7.42
Receptivity / ^{13}C	: 3.58
Gyromagnetic ratio	: 3.9366
Ouad. moment	: $-8\ 10^{-4}$

Resonance Freq. (MHz) referred to ^1H TMS resonance at 100 MHz

14.716_0

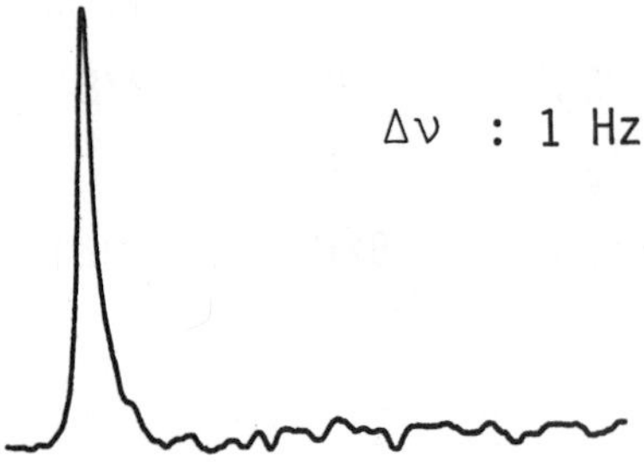

Δν : 1 Hz

Spectral width (Hz)	: 3 000
Number of data points	: 16 K
Pulse	: 20 μs/ 40°
Number of scans	: 100
Repetition rate (s)	: 7.7
^1H decoupled	: -
Exp. filter (Hz)	: 1

^7Li

Spin	: 3/2
Nat. abund (%)	: 92.58
Receptivity / ^{13}C	: $1.54\ 10^3$
Gyromagnetic ratio	: 10.3964
Ouad moment	: $-4.5\ 10^{-2}$

Resonance Freq. (MHz) referred to ^1H TMS resonance at 100 MHz

38.863_7

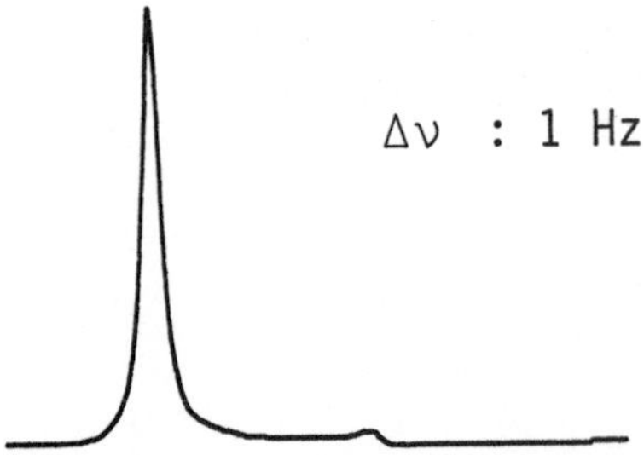

Δν : 1 Hz

Spectral width (Hz)	: 2 000
Number of data points	: 8 K
Pulse	: 13 μs/ 90°
Number of scans	: 1
Repetition rate (s)	: -
^1H decoupled	: -
Exp. filter (Hz)	: 1

CHEMICAL SHIFTS : reference LiCl M/D_2O

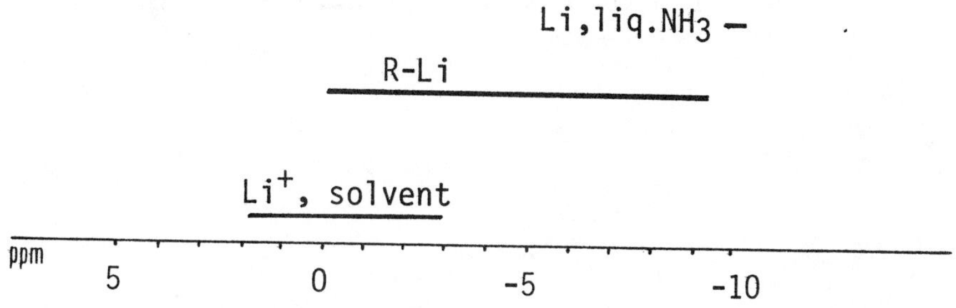

Li,liq.NH_3 —

R-Li

Li^+, solvent

ppm 5 0 -5 -10

COUPLING CONSTANTS : typical values (Hz)

	^1H	^{19}F	^{31}P	^{13}C	Homo
$\|^1J\|$ (^7Li)	- to -	- to -	- to -	10 to 45	< 0.5

RELAXATION :

T_1 typical values (s) : ^6Li : 10 to 80
^7Li : 3 to 0.3

LITTERATURE :

B.12 p. 129

REMARKS : ^6Li gives very sharp lines due to its low quadrupole moment

BERYLIUM Z 4

Sample	: $Be(NO_3)_2$	Lock	:	^2H
Solvent	: D_2O	Bo	:	5.875
Concentration	: 0.1 M	Temperature	:	27°

^9Be

Spin	: 3/2
Nat. abund (%)	: 100
Receptivity / ^{13}C	: 78.8
Gyromagnetic ratio	: 3.7589
Quad. moment	: $5.2.10^{-2}$

Resonance Freq. (MHz) referred to ^1H TMS resonance at 100MHz

14.051_8

$\Delta\nu$: 1,5 Hz

Spectral width (Hz)	: 1 500
Number of data points	: 8
Pulse	: 15 µs / 90°
Number of scans	: 1
Repetition rate (s)	: -
^1H decoupled	: -
Exp. filter (Hz)	: 1

CHEMICAL SHIFTS : reference $Be(NO_3)_2, H_2O$

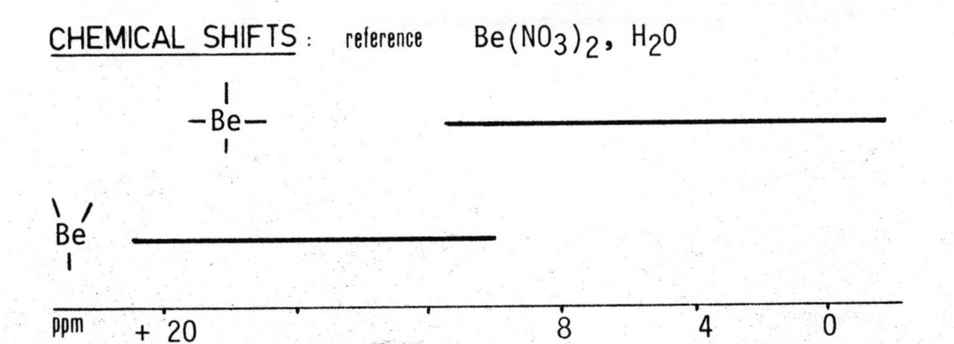

COUPLING CONSTANTS : typical values (Hz)

	^1H	^{19}F	^{31}P	^{13}C	Homo
$\| ^1J \|$	- to -	33 to	- to -	- to -	- to -

^1J BeP and ^1JBeH are too small to be measured

RELAXATION :

T_1 typical values (s) : 20 to 10^{-2}

LITTERATURE :

B. 12 p. 183

J.C. Kotz, R. Schaeffer and A. Clouse, Inorg. Chem., 6, 620 (1967)

R.A. Kovar and G.L. Morgan, J. Am. Chem. Soc., 92, 5067 (1970)

REMARKS :

BORON Z 5

Sample	: $NaBH_4$	Lock	: ^2H
Solvent	: D_2O	Bo	: 2.114
Concentration	: saturated	Temperature	: 27°

^{11}B

Spin	: 3/2
Nat abund (%)	: 80.42
Receptivity / ^{13}C	: 754
Gyromagnetic ratio	: 8.5794
Quad. moment	: 0.0355

Resonance Freq. (MHz) referred to ^1H TMS resonance at 100 MHz

32.071_5

$\Delta\nu$: 5 Hz

Spectral width (Hz)	: 2 000
Number of data points	: 4 K
Pulse	: 7 µs/ 50°
Number of scans	: 1
Repetition rate (s)	: -
^1H decoupled	: yes
Exp filter (Hz)	: 1

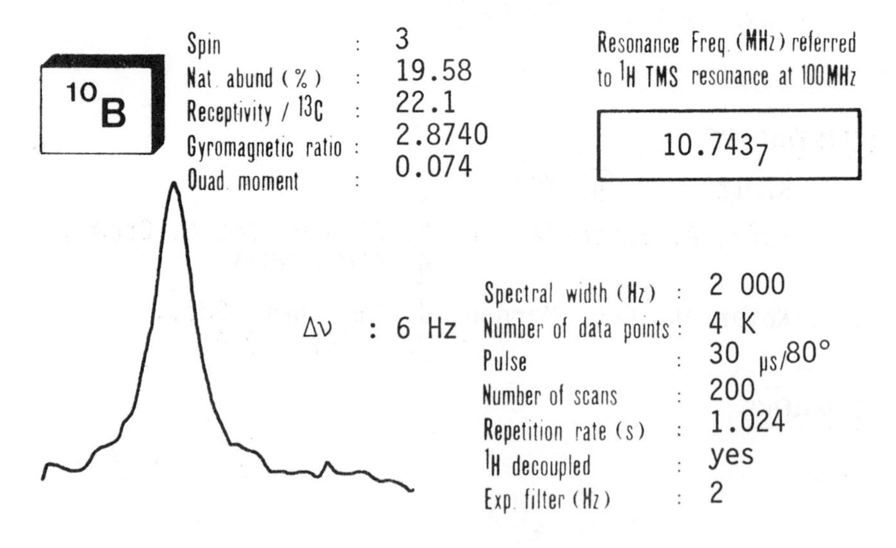

CHEMICAL SHIFTS : reference Et_2O, BF_3

Boranes

$>B<$

— $-B<$

ppm +40 0 -40

COUPLING CONSTANTS : typical values (Hz)

	^{1}H	^{19}F	^{31}P	^{13}C	Homo
$\lvert ^{1}J \rvert$ (^{11}B)	29 to 182	45 to 105	- to -	22 to 75	18.9

RELAXATION :

T_1 typical values (s) : 1 to 10^{-2}

LITTERATURE :

B. 10 Vol. 3 p. 87 (1969)

B. 24 (1978)

A.R. Garber et al., J. Magn. Res., 41, 458 (1980)

REMARKS :

A small quadrupole moment precludes the observation of fine structures.

NITROGEN Z 7

Sample	:	$MeNO_2$	Lock	:	^2H
Solvent	:	C_6D_6	Bo	:	5.875
Concentration	:	80/20(v/v)	Temperature	:	27°

^{14}N

Spin	:	1
Nat abund (%)	:	99.63
Receptivity / ^{13}C	:	5.69
Gyromagnetic ratio	:	1.9331
Quad. moment	:	$1.6\ 10^{-2}$

Resonance Freq. (MHz) referred to ^1H TMS resonance at 100 MHz

7.226_3

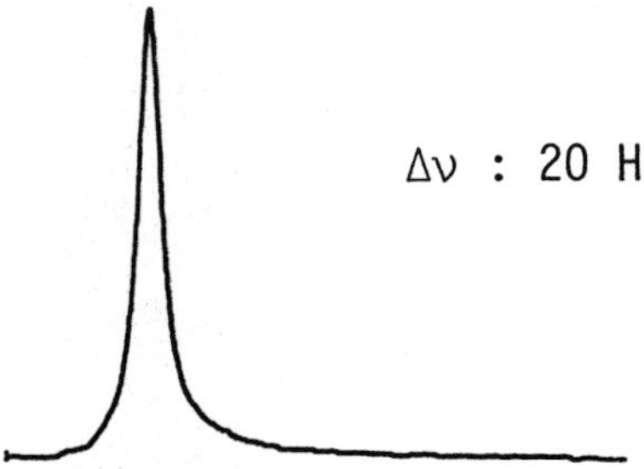

Δν : 20 Hz

Spectral width (Hz)	:	2 000
Number of data points	:	4 K
Pulse	:	63 μs/ 90°
Number of scans	:	1
Repetition rate (s)	:	-
^1H decoupled	:	-
Exp. filter (Hz)	:	5

^{15}N

Spin	:	1/2
Nat. abund (%)	:	0.37
Receptivity / ^{13}C	:	$2.19\ 10^{-2}$
Gyromagnetic ratio	:	-2.7116
Quad. moment	:	-

Resonance Freq. (MHz) referred to ^1H TMS resonance at 100 MHz

10.136_7

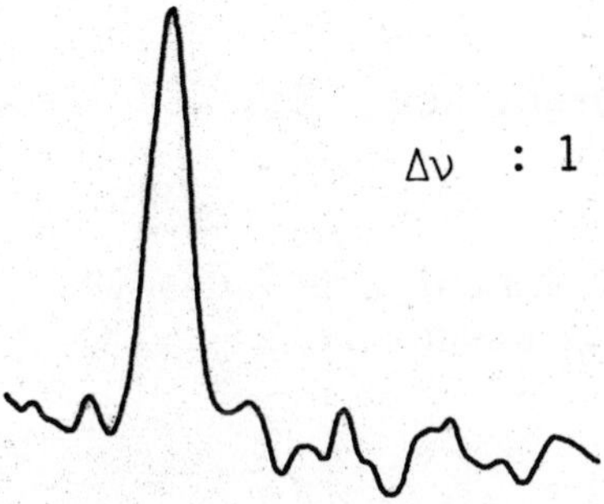

Δν : 1 Hz

Spectral width (Hz)	:	2 000
Number of data points	:	8 K
Pulse	:	20 μs/ 30°
Number of scans	:	110
Repetition rate (s)	:	7
^1H decoupled	:	yes
Exp. filter (Hz)	:	1

CHEMICAL SHIFTS: reference liquid NH_3 (25°)

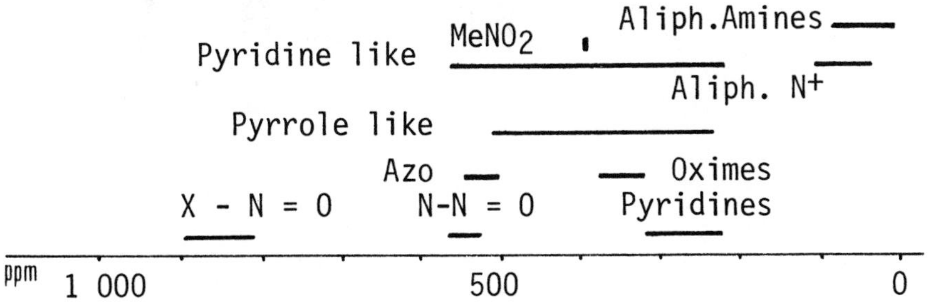

Pyridine like; $MeNO_2$; Aliph.Amines; Aliph. N^+

Pyrrole like

Azo; Oximes

X - N = 0; N-N = 0; Pyridines

ppm 1 000 ... 500 ... 0

COUPLING CONSTANTS: typical values (Hz)

	^1H	^{19}F	^{31}P	^{13}C	Homo
$\mid ^1J \mid$ (^{15}N)	60 to 140	180 to 210	0 to 100	0 to 36	4 to 25

RELAXATION:

T_1 typical values (s):

^{14}N : 0.3 to 3 10^{-3}

^{15}N : 0.5 to 170

LITTERATURE:

B 17

REMARKS: A protio/deutero (90/10) mixture of the appropriate solvent eases ^{15}N detection for non protonated nitrogens

OXYGEN Z 8

Sample	: D_2O	Lock	:	^2H
Solvent	: -	Bo	:	2.114
Concentration	: neat	Temperature	:	27°

^{17}O

Spin	:	5/2
Nat. abund (%)	:	0.037
Receptivity / ^{13}C	:	0.061
Gyromagnetic ratio	:	-3.6264
Quad. moment	:	-2.610-2

Resonance Freq. (MHz) referred to ^1H TMS resonance at 100MHz

13.556_4

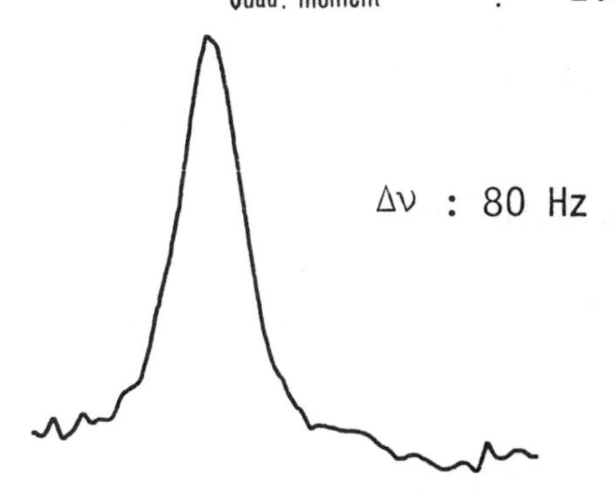

Δν : 80 Hz

Spectral width (Hz)	:	5 000
Number of data points	:	1 K
Pulse	:	30 μs / 90°
Number of scans	:	1 000
Repetition rate (s)	:	0.1
^1H decoupled	:	-
Exp. filter (Hz)	:	10

CHEMICAL SHIFTS : reference H_2O neat

$X_nO_m^{\pm}$

inorganic

organic -O-
O=

ppm + 1600 + 1200 + 800 + 400 0

COUPLING CONSTANTS : typical values (Hz)

	^1H	^{19}F	^{31}P	^{13}C	Homo
$\|^1J\|$	78 to 85	424	88 to 220	- to -	- to -

RELAXATION :

T_1 typical values (s) : 0.2 and below

LITTERATURE :

B. 12 p. 383

T.St Amour and D. Fiat, Bull. Magn. Reson., 1, 118 (1979)

W.G.K. Klemperer, Angew. Chem. Inst. Ed., 17, 246 (1978)

J.P. Kintzinger "Oxygen 17 and Silicon 29" (B8) Vol. 18 (1981)

REMARKS : ^{17}O is often observable, but high field is recommended.

SODIUM Z 11

Sample	: NaBr	Lock	: ^{2}H
Solvent	: D_2O	Bo	: 2.114
Concentration	: 9.8 M	Temperature	: 27°

^{23}Na

Spin	: 3/2
Nat. abund (%)	: 100
Receptivity / ^{13}C	: 525
Gyromagnetic ratio	: 7.0761
Quad. moment	: 0.12

Resonance Freq. [MHz] referred to ^{1}H TMS resonance at 100MHz

26.451_9

$\Delta\nu$: 14 Hz

Spectral width (Hz)	: 1 000
Number of data points	: 4 K
Pulse	: 2 μs / 20°
Number of scans	: 1
Repetition rate (s)	: -
^{1}H decoupled	: -
Exp. filter (Hz)	: 1

CHEMICAL SHIFTS : reference $Na^+Cl(1\ m)/H_2O$

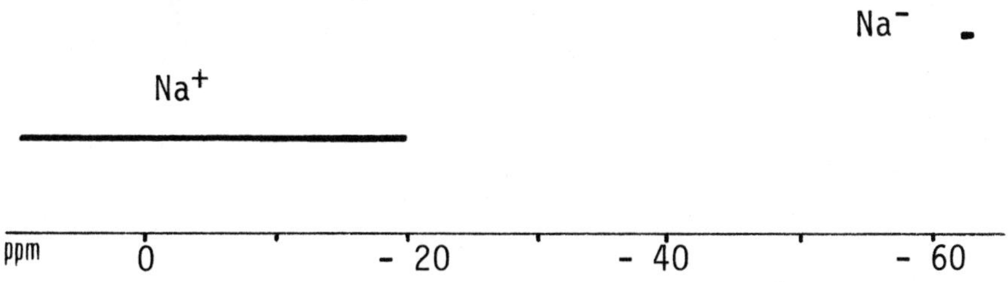

Na$^-$

Na$^+$

ppm 0 - 20 - 40 - 60

COUPLING CONSTANTS : typical values (Hz)

	^1H	^{19}F	^{31}P	^{13}C	Homo
^1J	- to -	- to -	- to -	- to -	- to -

RELAXATION :

T_1 typical values (s) : 0.06 and below

LITTERATURE :

B. 12 p. 129

S. Forsen and B. Lindman in "Methods of biochemical analysis", Vol. 27 (1981), John Wiley

P. Laszlo, Angew. Chem. Inst.Ed., 17, 254 (1978)

REMARKS : Easily observed. This nucleus is of great interest in biological studies

MAGNESIUM Z 12

Sample	: $MgCl_2$	Lock	:	^2H
Solvent	: D_2O	Bo	:	2.114
Concentration	: 9.88 M	Temperature	:	27°

^{25}Mg

Spin	:	5/2
Nat. abund (%)	:	10.13
Receptivity / ^{13}C	:	1.54
Gyromagnetic ratio	:	1.6375
Quad. moment	:	0.22

Resonance Freq. (MHz) referred to ^1H TMS resonance at 100MHz

6.121_6

$\Delta\nu$: 7 Hz

Spectral width (Hz)	:	2 000
Number of data points	:	4 K
Pulse	:	30 μs / 70°
Number of scans	:	1 000
Repetition rate (s)	:	1 024
^1H decoupled	:	-
Exp. filter (Hz)	:	2

CHEMICAL SHIFTS : reference $MgCl_2/H_2O$

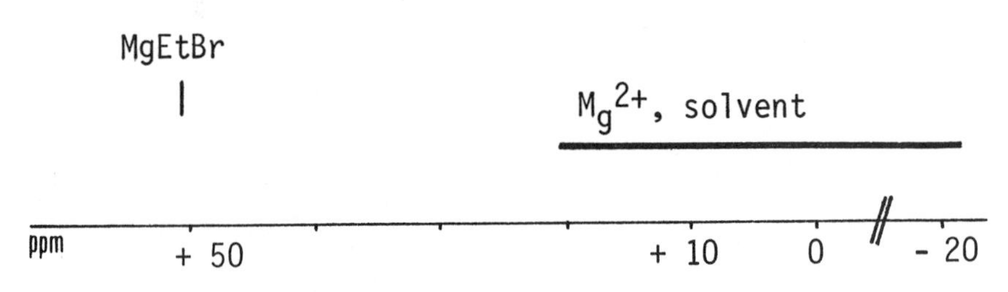

COUPLING CONSTANTS : typical values (Hz)

	^1H	^{19}F	^{31}P	^{13}C	Homo
^1J	- to -	- to -	- to -	- to -	- to -

RELAXATION :

T_1 typical values (s) : below 0.22

LITTERATURE :

B. 12 p. 183

- S. Forsen and B. Lindman in "Methods of Biochemical Analysis", Vol. 27 (1981) John Wiley
- S. Forsen and B. Lindman B 21 Vol. 11 (1981)

REMARKS : Millimolar solutions may be observed for symmetrical species as Mg^{2+}.

ALUMINIUM Z 13

Sample	: $Al(NO_3)_3$	Lock	:	^2H
Solvent	: D_2O	Bo	:	2.114
Concentration	: 1.5 M	Temperature	:	27°

^{27}Al

Spin	:	5/2
Nat. abund (%)	:	100
Receptivity / ^{13}C	:	1.17 10^2
Gyromagnetic ratio	:	6.9704
Quad. moment	:	0.149

Resonance Freq. (MHz) referred to ^1H TMS resonance at 100MHz

26.056_8

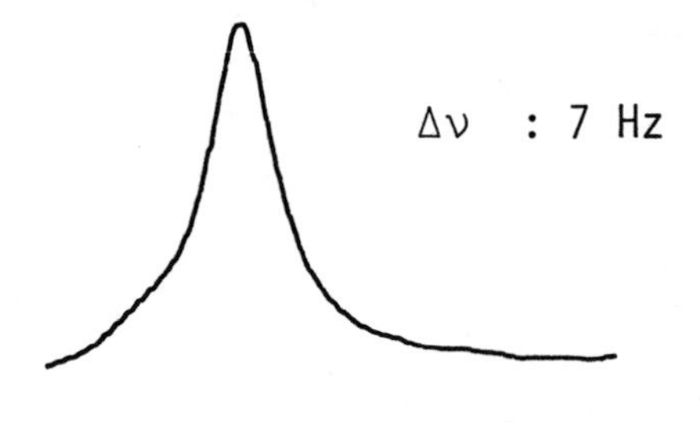

Δν : 7 Hz

Spectral width (Hz)	:	2 000
Number of data points	:	4 K
Pulse	:	4 μs / 20°
Number of scans	:	1
Repetition rate (s)	:	-
^1H decoupled	:	-
Exp. filter (Hz)	:	1

CHEMICAL SHIFTS : reference $Al(H_2O)_6^{3+}$

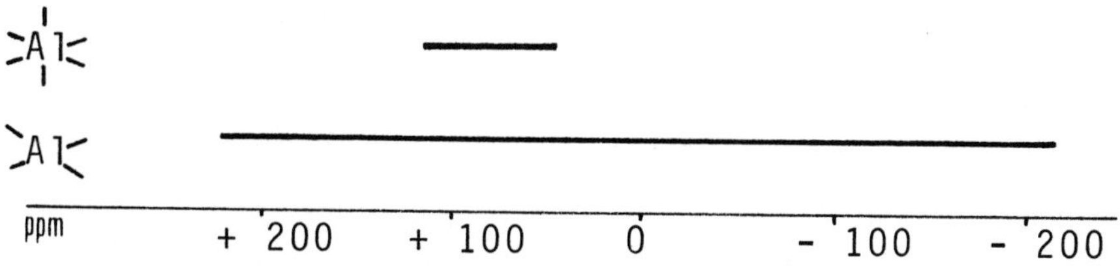

COUPLING CONSTANTS : typical values (Hz)

	^1H	^{19}F	^{31}P	^{13}C	Homo
$\|^1J\|$	110 to 185	- to -	240 to 290	71 to 110	- to -

RELAXATION :

T_1 typical values (s) : 1 to below 10^{-3}

LITTERATURE :

B 12 p. 279

B 21 - Vol. 5 A - p. 465

REMARKS : Easy to observe in symmetrical environment

SILICON Z 14

Sample	: TMS	Lock	: C_6F_6 cap.
Solvent	: C_6D_6	Bo	: 2.114
Concentration	: 4.2 M	Temperature	: 27°

^{29}Si

Spin	: 1/2
Nat. abund (%)	: 4.70
Receptivity / ^{13}C	: 2.09
Gyromagnetic ratio	: -5.3146
Quad. moment	: -

Resonance Freq. (MHz) referred to ^1H TMS resonance at 100MHz

19.867_1

Δν : 0.6 Hz

Spectral width (Hz)	: 400
Number of data points	: 8 K
Pulse	: 15 μs / 80°
Number of scans	: 25
Repetition rate (s)	: 5.12
^1H decoupled	: yes
Exp. filter (Hz)	: 0.2

CHEMICAL SHIFTS : reference $Si(Me_4)$ neat

SiX_nY_{4-n}

$SiH_n R_{4-n}$

$Si_n X_{2n+2}$

Siloxanes

ppm + 100 0 - 100

COUPLING CONSTANTS : typical values (Hz)

	^1H	^{19}F	^{31}P	^{13}C	Homo
$\lvert ^1J \rvert$	147 to 420	110 to 488	7 to 50	37 to 113	52 to 186

RELAXATION :

T_1 typical values (s) : 150 to 5

LITTERATURE :

B. 12 p. 310

J. Schraml et al., in "Determination of Organic Structures by Physical Methods", Vol. 6, ch.4 (1976)

B. 8, "Oxygen 17 and Silicon 29", Vol. 18 (1981)

REMARKS : Due to negative NOE the use of relaxagents or gated decoupling is often necessary.
Insert or tube give a broad line (-80 to -300 ppm)

PHOSPHORUS Z 15

Sample	: H_3PO_4 85 %	Lock	: ^2H
Solvent	: D_2O	Bo	: 2.114
Concentration	: 19.8 M	Temperature	: 27 °

^{31}P

Spin	: 1/2
Nat. abund (%)	: 100
Receptivity / ^{13}C	: 377
Gyromagnetic ratio	: 10.8289
Quad. moment	: -

Resonance Freq. (MHz) referred to ^1H TMS resonance at 100MHz

40.480_7

$\Delta\nu$: 2.8 Hz

Spectral width (Hz)	: 600
Number of data points	: 8 K
Pulse	: 16 µs / 90°
Number of scans	: 1
Repetition rate (s)	: -
^1H decoupled	: -
Exp. filter (Hz)	: -

CHEMICAL SHIFTS : reference H_3PO_4 85 % H_2O

P(V)

P(IV)

P(III)

P(II)

ppm + 200 0 - 200

COUPLING CONSTANTS : typical values (Hz)

	^1H	^{19}F	^{31}P	^{13}C	Homo
$\mid ^1J \mid$	37 to 1100	108 to 1441	See Homo	50 to 304	443 to 590

RELAXATION :

T_1 typical values (s) : 55 to 10^{-1}

LITTERATURE :

B. 10 Vol. 6 (1971)
B. 21 Vol. 5 B (1973)

REMARKS :

SULFUR Z 16

Sample	: $(NH_4)_2SO_4$	Lock	: ^2H
Solvent	: D_2O	Bo	: 2.114
Concentration	: saturated	Temperature	: 27°

^{33}S

Spin	: 3/2
Nat. abund (%)	: 0.76
Receptivity / ^{13}C	: $9.73 10^{-2}$
Gyromagnetic ratio	: 2.0534
Quad. moment	: -0.05

Resonance Freq. (MHz) referred to ^1H TMS resonance at 100MHz

7.676_0

$\Delta\nu$: 7 Hz

Spectral width (Hz)	: 2 000
Number of data points	: 1 K
Pulse	: 30 µs / 90°
Number of scans	: 5 000
Repetition rate (s)	: 0.26
^1H decoupled	: -
Exp. filter (Hz)	: 4

CHEMICAL SHIFTS : reference CS_2 neat

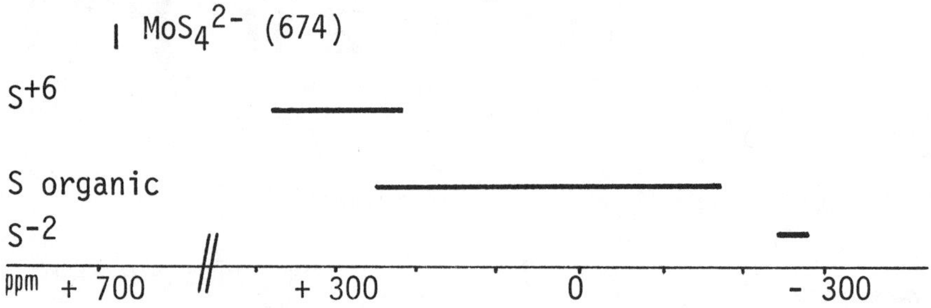

| MoS_4^{2-} (674)

S^{+6}

S organic

S^{-2}

ppm + 700 // + 300 0 - 300

COUPLING CONSTANTS : typical values (Hz)

	^1H	^{19}F	^{31}P	^{13}C	Homo
$\|^1J\|$	- to -	251 (SF_6)	- to -	- to -	- to -

RELAXATION :

T_1 typical values (s) : 10^{-2} to < 10^{-3}

LITTERATURE :

B. 12 p. 401

P. Kroneck, O. Lutz and A. Nolle : Z. Naturforsch., 35 a, 226 (1979)

REMARKS : very low receptivity, broad lines. Symmetric environnement is required for easy detection

CHLORINE Z 17

Sample	: KCl	Lock	: ^2H
Solvent	: D_2O	Bo	: 2.114
Concentration	: 2.21 M	Temperature	: 27°

^{35}Cl

Spin	: 3/2
Nat abund (%)	: 75.53
Receptivity / ^{13}C	: 20.2
Gyromagnetic ratio	: 2.6210
Quad. moment	: $-8.0\ 10^{-2}$

Resonance Freq. (MHz) referred to ^1H TMS resonance at 100 MHz

9.797_9

$\Delta\nu$: 17 Hz

Spectral width (Hz)	: 2 000
Number of data points	: 4 K
Pulse	: 30 µs/90°
Number of scans	: 7
Repetition rate (s)	: 1.024
^1H decoupled	: -
Exp. filter (Hz)	: 4

^{37}Cl

Spin	: 3/2
Nat. abund (%)	: 24.47
Receptivity / ^{13}C	: 3.8
Gyromagnetic ratio	: 2.1817
Quad. moment	: $-6.32\ 10^{-2}$

Resonance Freq. (MHz) referred to ^1H TMS resonance at 100 MHz

8.155_7

$\Delta\nu$: 12 Hz

Spectral width (Hz)	: 2 000
Number of data points	: 4 K
Pulse	: 30 µs/ 80°
Number of scans	: 139
Repetition rate (s)	: 1.024
^1H decoupled	: -
Exp. filter (Hz)	: 4

CHEMICAL SHIFTS : reference Cl^- (NaCl in H_2O)

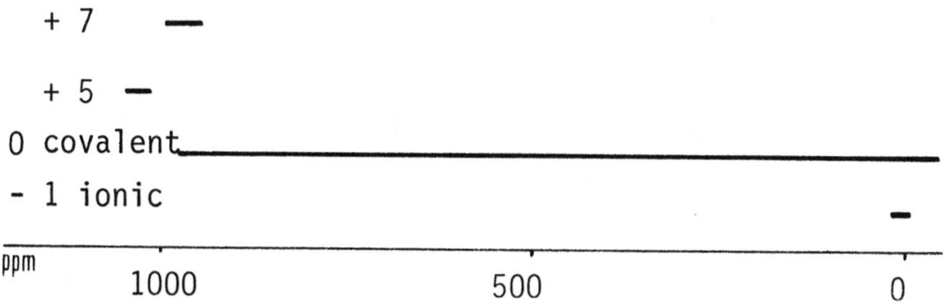

+ 7

+ 5

0 covalent

- 1 ionic

ppm 1000 500 0

COUPLING CONSTANTS : typical values (Hz)

	^1H	^{19}F	^{31}P	^{13}C	Homo
$\lvert ^1J \rvert$ (^{35}Cl)	41 (HCl)	20 to 340	105 to 120	23 ($CHCl_3$)	- to -

RELAXATION :

T_1 typical values (s) :

Covalent : below 10^{-4}

ionic : 0.2 to 10^{-4}

LITTERATURE :

B. 8, Vol. 12

REMARKS : Chlorine quadrupole moment will generaly preclude direct high resolution NMR observation of covalent chlorine compounds (broad lines)

POTASSIUM Z 19

Sample	:	K Br	Lock	:	^2H
Solvent	:	D_2O	Bo	:	5.875
Concentration	:	Saturated	Temperature	:	27°

^{39}K

Spin	:	3/2
Nat abund (%)	:	93.1
Receptivity / ^{13}C	:	2.69
Gyromagnetic ratio	:	1.2483
Quad. moment	:	$5.5\ 10^{-2}$

Resonance Freq. (MHz) referred to ^1H TMS resonance at 100 MHz

4.666_4

$\Delta\nu$: 13 Hz

Spectral width (Hz)	:	2 000
Number of data points	:	4 K
Pulse	:	80 μs/ 90°
Number of scans	:	12
Repetition rate (s)	:	1
^1H decoupled	:	-
Exp. filter (Hz)	:	5

^{41}K

Spin	:	3/2
Nat. abund (%)	:	6.88
Receptivity / ^{13}C	:	$3.28\ 10^{-2}$
Gyromagnetic ratio	:	0.6851
Quad. moment	:	$6.7\ 10^{-2}$

Resonance Freq. (MHz) referred to ^1H TMS resonance at 100 MHz

2.561_3

$\Delta\nu$: 16 Hz

Spectral width (Hz)	:	2 500
Number of data points	:	4 K
Pulse	:	80 μs/ 90°
Number of scans	:	2 000
Repetition rate (s)	:	0,82
^1H decoupled	:	-
Exp. filter (Hz)	:	5

CHEMICAL SHIFTS : reference K^+ - infinite dilution

^-K may increase the range to about - 150 ppm

K^+, solvent,

ppm + 10 0 - 10

COUPLING CONSTANTS : typical values (Hz)

	^1H	^{19}F	^{31}P	^{13}C	Homo
^1J	- to -	- to -	- to -	- to -	- to -

RELAXATION :

T_1 typical values (s) : $2.\ 10^{-2}$

LITTERATURE :

B. 12 p. 129

S. Forsen and B. Lindman in "Methods of Biochemical Analysis", Vol. 27, (1981), John Wiley

REMARKS : ^{39}K is the generally observed isotope. Potassium has the lowest sensitivity among the Alkali metals.

CALCIUM Z 20

Sample	: $CaCl_2$	Lock	: ^2H
Solvent	: D_2O	Bo	: 5.875
Concentration	: Saturated	Temperature	: 27°

^{43}Ca

Spin	: 7/2
Nat. abund (%)	: 0.145
Receptivity / ^{13}C	: $5.27\ 10^{-2}$
Gyromagnetic ratio	: - 1.8001
Quad. moment	: - 0.05

Resonance Freq. (MHz) referred to ^1H TMS resonance at 100MHz

6.7294

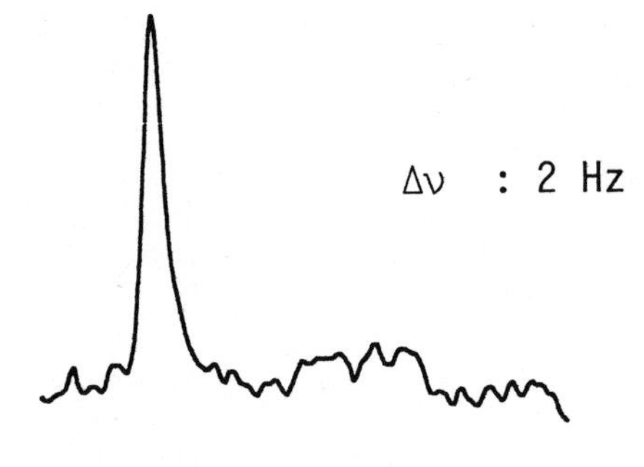

$\Delta\nu$: 2 Hz

Spectral width (Hz)	: 3 000
Number of data points	: 8 K
Pulse	: 30 μs / 90°
Number of scans	: 500
Repetition rate (s)	: 1.35
^1H decoupled	: -
Exp. filter (Hz)	: 1

CHEMICAL SHIFTS: reference $CaCl_2$, aqueous

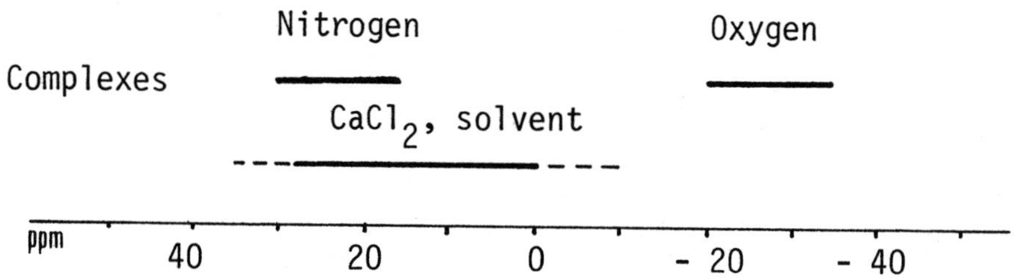

Nitrogen Oxygen

Complexes

$CaCl_2$, solvent

ppm 40 20 0 - 20 - 40

COUPLING CONSTANTS: typical values (Hz)

	^1H	^{19}F	^{31}P	^{13}C	Homo
^1J	- to -	- to -	- to -	- to -	- to -

RELAXATION :

T_1 typical values (s) : 1.33 and below

LITTERATURE :

B 12 p. 183

S. Forsen and B. Lindman: "Methods of Biochemical Analysis", Vol. 27 (1981), John Wiley

S. Forsen and B. Lindman, B 21 Vol.11 (1981)

REMARKS: FT NMR allows to study millimolar solutions with enriched Ca^{2+} compounds

SCANDIUM Z 21

Sample	:	$ScCl_3$	Lock	:	^2H
Solvent	:	D_2O	Bo	:	5.875
Concentration	:	1 M	Temperature	:	27°

^{45}Sc

Spin	:	7/2
Nat. abund (%)	:	100
Receptivity / ^{13}C	:	$1.71\ 10^3$
Gyromagnetic ratio	:	6.4982
Quad. moment	:	- 0,22

Resonance Freq. (MHz) referred to ^1H TMS resonance at 100MHz

24.291_7

$\Delta\nu$: 273 Hz

Spectral width (Hz)	:	10.000
Number of data points	:	4 K
Pulse	:	40 μs / 90°
Number of scans	:	1
Repetition rate (s)	:	-
^1H decoupled	:	-
Exp. filter (Hz)	:	50

CHEMICAL SHIFTS : reference $ScCl_3$. M/D_2O

$(ScX_6)^{n-}$

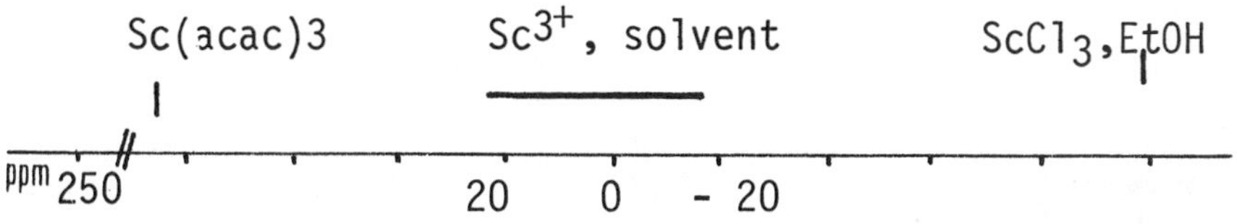

Sc(acac)3 | Sc^{3+}, solvent | $ScCl_3$,EtOH

ppm 250 // ... 20 0 - 20

COUPLING CONSTANTS : typical values (Hz)

	^1H	^{19}F	^{31}P	^{13}C	Homo
$\lvert ^1J \rvert$	- to -	180 (ScF_6^{3-})	- to -	- to -	- to -

RELAXATION :

T_1 typical values (s) : 3. 10^{-3} to 10^{-4}

LITTERATURE :

B 12 p. 196

B 21 Vol. 10 A (1980)

Y.A. Buslaev et al., Koordi. Khim. 6, 361 (1980)

REMARKS :

TITANIUM Z 22

Sample	: $TiCl_4$	Lock	:	unlocked
Solvent	: neat	Bo	:	5.875
Concentration	: -	Temperature	:	27°

^{47}Ti

Spin	:	5/2
Nat abund (%)	:	7.28
Receptivity / ^{13}C	:	0.864
Gyromagnetic ratio	:	1.5084
Quad. moment	:	+ 0.29

Resonance Freq. (MHz) referred to ^{1}H TMS resonance at 100 MHz

5.639_0

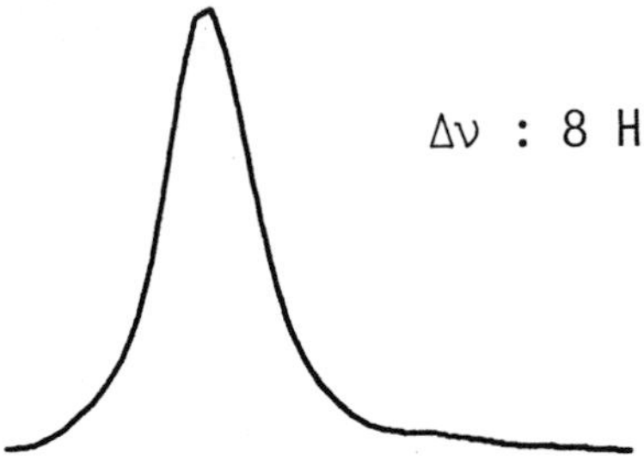

$\Delta\nu$: 8 Hz

Spectral width (Hz)	:	10 000
Number of data points	:	8 K
Pulse	:	100 μs/ 90°
Number of scans	:	48
Repetition rate (s)	:	5.4
^{1}H decoupled	:	-
Exp. filter (Hz)	:	1

^{49}Ti

Spin	:	7/2
Nat. abund (%)	:	5.51
Receptivity / ^{13}C	:	1.18
Gyromagnetic ratio	:	1.5080
Quad. moment	:	+ 0.24

Resonance Freq. (MHz) referred to ^{1}H TMS resonance at 100 MHz

5.637_5

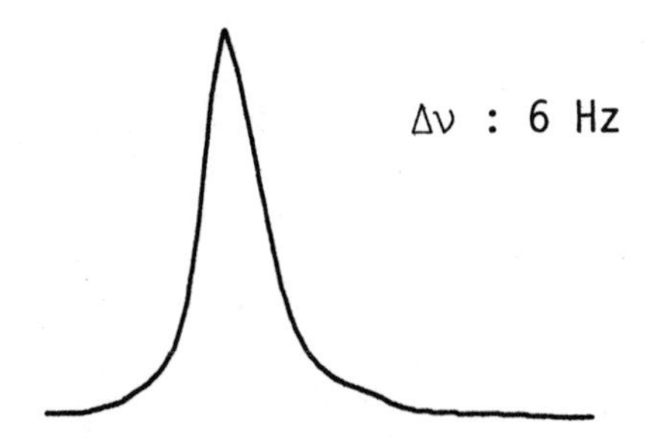

$\Delta\nu$: 6 Hz

Spectral width (Hz)	:	10 000
Number of data points	:	8 K
Pulse	:	100 μs/ 90°
Number of scans	:	48
Repetition rate (s)	:	5.4
^{1}H decoupled	:	-
Exp. filter (Hz)	:	1

CHEMICAL SHIFTS : reference $(TiF_6)^{2-}$ / 48 % HF

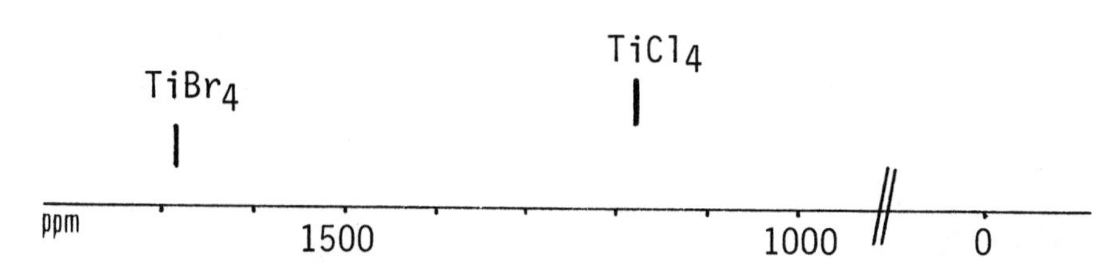

COUPLING CONSTANTS : typical values (Hz)

	^1H	^{19}F	^{31}P	^{13}C	Homo
$\lvert ^1J \rvert$	- to -	33 $(TiF_6)^{2-}$	- to -	- to -	- to -

RELAXATION :

T_1 typical values (s) : 10^{-2} to 10^{-4}

LITTERATURE :

B. 12 p. 201
B. 21 Vol. 10 A (1980)

REMARKS : ^{47}Ti and ^{49}Ti resonances are always found ca.270 ppm apart

VANADIUM Z 23

Sample	: $VOCl_3$		Lock	: ^2H
Solvent	: C_6D_6		Bo	: 2.114
Concentration	: 90/10 (v/v)		Temperature	: 27°

^{51}V

Spin	:	7/2
Nat abund (%)	:	99.76
Receptivity / ^{13}C	:	2150
Gyromagnetic ratio	:	7.0362
Quad. moment	:	- 0.052

Resonance Freq. (MHz) referred to ^1H TMS resonance at 100 MHz

26.302_9

$\Delta\nu$: 26 Hz

Spectral width (Hz)	:	1 000
Number of data points	:	4 K
Pulse	:	0.5 μs/3°
Number of scans	:	1
Repetition rate (s)	:	-
^1H decoupled	:	-
Exp. filter (Hz)	:	4

^{50}V

Spin	:	6
Nat. abund (%)	:	0.24
Receptivity / ^{13}C	:	0.755
Gyromagnetic ratio	:	2.6491
Quad. moment	:	+ 0.21

Resonance Freq. (MHz) referred to ^1H TMS resonance at 100 MHz

9.970_3

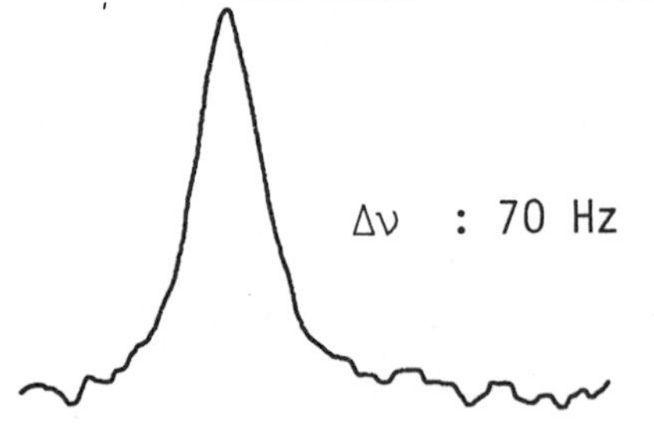

$\Delta\nu$: 70 Hz

Spectral width (Hz)	:	2 000
Number of data points	:	4 K
Pulse	:	30 μs/ 90°
Number of scans	:	1 023
Repetition rate (s)	:	1 024
^1H decoupled	:	-
Exp. filter (Hz)	:	10

CHEMICAL SHIFTS : reference $VOCl_3$ neat

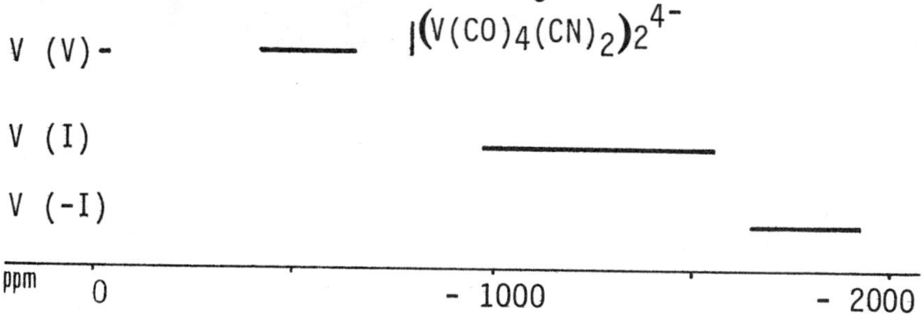

V (V)- ——— $[(V(CO)_4(CN)_2)_2]^{4-}$

V (I) ———

V (-I) ———

ppm 0 - 1000 - 2000

COUPLING CONSTANTS : typical values (Hz)

	^1H	^{19}F	^{31}P	^{13}C	Homo
$\|^1J\|$ (^{51}V)	- to -	88 to 140	160 to 660	116	- to -

RELAXATION :

T_1 typical values (s) : below 5 10^{-2}

LITTERATURE :

B. 12 p. 203

R. Kidd "Nuclear Shielding of the Transitions Metals", B.21 - Vol. 10 A (1980)

REMARKS : ^{51}V is easy to observe. δ and J are valuable tools for structures determinations

CHROMIUM Z 24

Sample	:	$CrO_4(NH_4)_2$	Lock	:	^2H
Solvent	:	D_2O	Bo	:	5.875
Concentration	:	1 M	Temperature	:	27°

^{53}Cr

Spin	:	3/2
Nat. abund (%)	:	9.55
Receptivity / ^{13}C	:	0.49
Gyromagnetic ratio	:	- 1.5120
Quad. moment	:	± 0.03

Resonance Freq. (MHz) referred to ^1H TMS resonance at 100MHz

5.6524

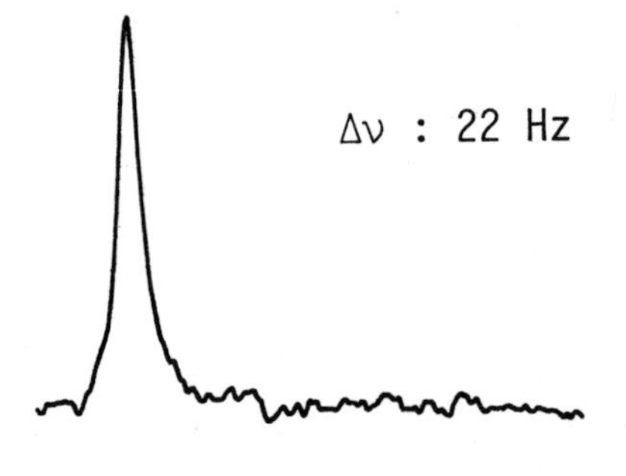

$\Delta\nu$: 22 Hz

Spectral width (Hz)	:	6 000
Number of data points	:	4 K
Pulse	:	100 μs / 90°
Number of scans	:	2 000
Repetition rate (s)	:	0.34
^1H decoupled	:	-
Exp. filter (Hz)	:	4

CHEMICAL SHIFTS : reference CrO_4^{2-}, aqueous

$Cr(CO)_6$ (- 1795)

ppm 500 0 - 500

COUPLING CONSTANTS : typical values (Hz)

	^1H	^{19}F	^{31}P	^{13}C	Homo
^1J	- to -	- to -	- to -	- to -	- to -

RELAXATION :

T_1 typical values (s) : 10^{-2} in a symetrical environment

LITTERATURE :

B 12 p. 212

B 21 Vol. 10 A (1980)

REMARKS : $^1J_{Cr-^{17}O}$ = 10 ± 2 Hz in CrO_4^{2-}

MANGANESE Z 25

Sample	: $KMnO_4$	Lock	:	^2H
Solvent	: D_2O	Bo	:	2.114
Concentration	: saturated	Temperature	:	27°

^{55}Mn

Spin	:	5/2
Nat. abund (%)	:	100
Receptivity / ^{13}C	:	994
Gyromagnetic ratio	:	6.6195
Quad. moment	:	0.55

Resonance Freq. (MHz) referred to ^1H TMS resonance at 100MHz

24.745_0

$\Delta\nu$: 8 Hz

Spectral width (Hz)	:	5 000
Number of data points	:	2 K
Pulse	:	15 μs / 80°
Number of scans	:	250
Repetition rate (s)	:	0.2
^1H decoupled	:	-
Exp. filter (Hz)	:	1

CHEMICAL SHIFTS : reference $KMnO_4$ / H_2O

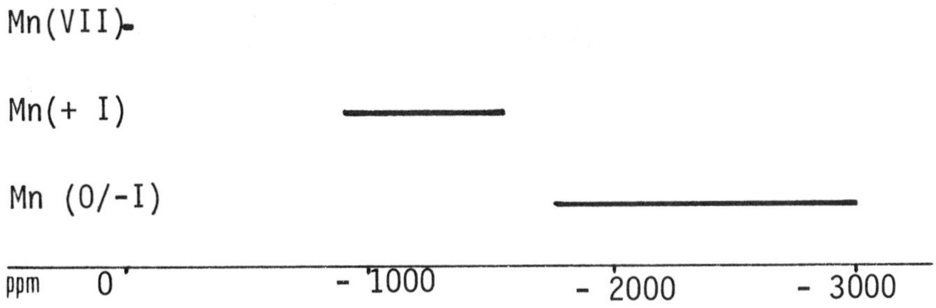

Mn(VII) -

Mn(+ I) ——

Mn (0/-I) ——

ppm 0 - 1000 - 2000 - 3000

COUPLING CONSTANTS : typical values (Hz)

	^1H	^{19}F	^{31}P	^{13}C	Homo
^1J	- to -	- to -	- to -	- to -	- to -

only $^1J_{Mn\ 17O}$ = 29 Hz is known

RELAXATION :

T_1 typical values (s) : 0.8 and below

LITTERATURE :

B. 12 p. 218

R.G. Kidd, et al., Inorg.Chem., 12, 728 (1973)

R.G. Kidd "Nuclear Shielding of the Transition Metals", (B 21), Vol. 10 A, (1980)

REMARKS : Only Mn(1, 0, +1, +7) may be observed.
Solvent effect of 40 ppm is sometimes found.

IRON Z 26

Sample	: $Fe(CO)_5$	Lock	:	^2H
Solvent	: C_6D_6	Bo	:	5.875
Concentration	: 80/20 (v/v)	Temperature	:	27°

^{57}Fe

Spin	:	1/2
Nat. abund (%)	:	2.19
Receptivity / ^{13}C	:	4.2 10^{-3}
Gyromagnetic ratio	:	0.8661
Quad. moment	:	-

Resonance Freq. (MHz) referred to ^1H TMS resonance at 100MHz

3.2377

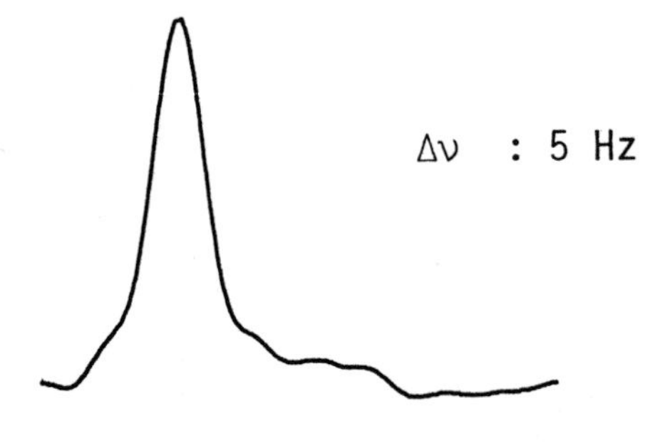

Spectral width (Hz)	:	10 000
Number of data points	:	1 K
Pulse	:	60 μs / 45°
Number of scans	:	45.10^4
Repetition rate (s)	:	0.17
^1H decoupled	:	-
Exp. filter (Hz)	:	5

CHEMICAL SHIFTS : reference $Fe(CO)_5$

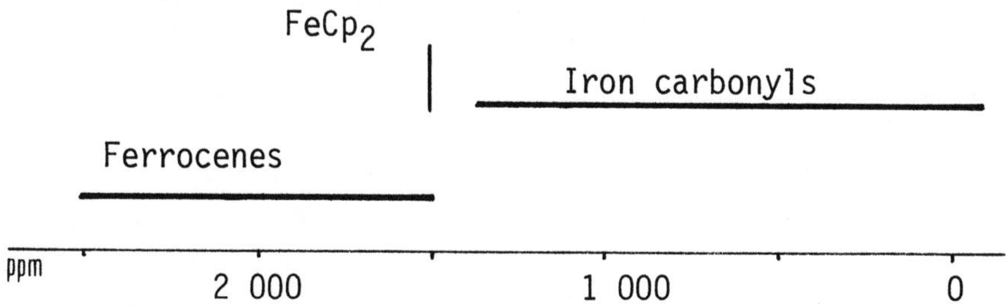

FeCp$_2$

Iron carbonyls

Ferrocenes

ppm 2 000 1 000 0

COUPLING CONSTANTS : typical values (Hz)

	^1H	^{19}F	^{31}P	^{13}C	Homo
$\|^1J\|$	- to -	- to -	25 to 28	23 to 32	- to -

RELAXATION :

T_1 typical values (s) : above 10

LITTERATURE :

B. 12 p. 224 ; B 21, Vol. 10 A (1980)

Von Philipsborn et al. : J. Organometal. Chem., 1980 (to be published)

REMARKS : ^{57}Fe spectra giving a single line can be accumulated under very fast pulsing rate J. Magn. Res. 5, vol 3, 376 (1971)

COBALT Z 27

Sample	: $K_3Co(CN)_6$	Lock	: ^2H
Solvent	: D_2O	Bo	: 2.114
Concentration	: saturated	Temperature	: 27°

^{59}Co

Spin	: 7/2
Nat. abund (%)	: 100
Receptivity / ^{13}C	: 1570
Gyromagnetic ratio	: 6.3472
Quad. moment	: 0.40

Resonance Freq. (MHz) referred to ^1H TMS resonance at 100MHz

23.727_1

$\Delta\nu$: 5 Hz

Spectral width (Hz)	: 600
Number of data points	: 8 K
Pulse	: 3 μs / 15°
Number of scans	: 1
Repetition rate (s)	: -
^1H decoupled	: -
Exp. filter (Hz)	: 0.5

CHEMICAL SHIFTS : reference $K_3Co(CN)_6$

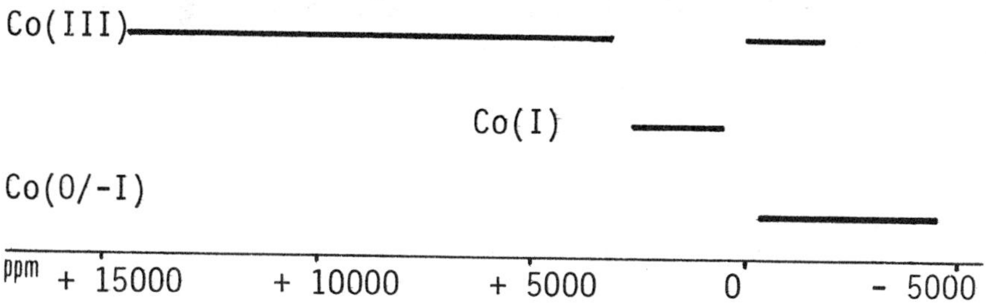

Co(III)

Co(I)

Co(0/-I)

ppm + 15000 + 10000 + 5000 0 - 5000

COUPLING CONSTANTS : typical values (Hz)

	^1H	^{19}F	^{31}P	^{13}C	Homo
$\|^1J\|$	- to -	- to -	414 to 1227	125 to 287	- to -

RELAXATION :

T_1 typical values (s) : 0.1 and below

LITTERATURE :

B. 12 p. 225

R.G. Kidd, "Nuclear Shielding of the Transition Metals", B. 21, Vol. 10 A, (1980)

Y. Yamasaki, Report of the University of Electrocommunication, 27, 291 (1977) and 29, 69 (1978)

REMARKS : Only Co III, I, 0, -I lead to high resolution NMR. Small concentration effect < 20 ppm. Large solvent effect (100 ppm). Very large temperature effect (- 1.5 to - 3 ppm/°K).

NICKEL Z 28

Sample	: $Ni(CO)_4$		Lock	: ^2H
Solvent	: C_6D_6		Bo	: 5.875
Concentration	: 80/20 (v/v)		Temperature	: 27°

^{61}Ni

Spin	: 3/2
Nat. abund (%)	: 1.19
Receptivity / ^{13}C	: 0.24
Gyromagnetic ratio	: -2.3904
Quad. moment	: (-)

Resonance Freq. (MHz) referred to ^1H TMS resonance at 100MHz

8.936_1

$\Delta\nu$: 7 Hz

Spectral width (Hz)	: 1 500
Number of data points	: 2 K
Pulse	: 70 µs / 90°
Number of scans	: 100
Repetition rate (s)	: 0.68
^1H decoupled	: -
Exp. filter (Hz)	: 1

CHEMICAL SHIFTS : reference $Ni(Co)_4$ in C_6D_6

$Ni(Co)_3L$

ppm 50 10 0

COUPLING CONSTANTS : typical values (Hz)

	^1H	^{19}F	^{31}P	^{13}C	Homo
$\lvert ^1J \rvert$	- to -	- to -	150 to 210	- to -	- to -

RELAXATION :

T_1 typical values (s) : 10^{-2} in symmetrical environment

LITTERATURE : B. 21, Vol. 10 A (1980)

H. Schumann et al. : Z. Naturforsch. 35 b 639 (1980)

REMARKS : Caution : $Ni(Co)_4$ is one of the most toxic chemicals

COPPER Z 29

Sample	:	CuCN	Lock	:	^2H
Solvent	:	D_2O, KCN sat	Bo	:	2.114
Concentration	:	half sat.	Temperature	:	27°

^{63}Cu

Spin	:	3/2
Nat abund (%)	:	69.09
Receptivity / ^{13}C	:	365
Gyromagnetic ratio	:	7.0965
Quad. moment	:	- 0.211

Resonance Freq. (MHz) referred to ^1H TMS resonance at 100 MHz

26.528_1

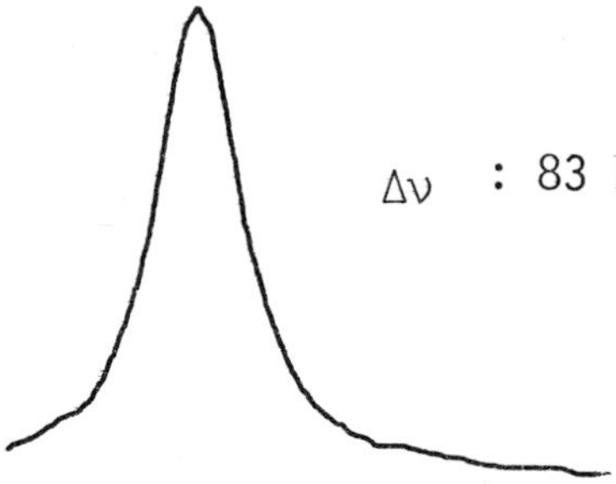

$\Delta\nu$: 83 Hz

Spectral width (Hz)	:	5 000
Number of data points	:	1 K
Pulse	:	13 μs/90°
Number of scans	:	50
Repetition rate (s)	:	0.1
^1H decoupled	:	-
Exp. filter (Hz)	:	10

^{65}Cu

Spin	:	3/2
Nat abund (%)	:	30.91
Receptivity / ^{13}C	:	201
Gyromagnetic ratio	:	7.6018
Quad. moment	:	- 0.195

Resonance Freq. (MHz) referred to ^1H TMS resonance at 100 MHz

28.417_2

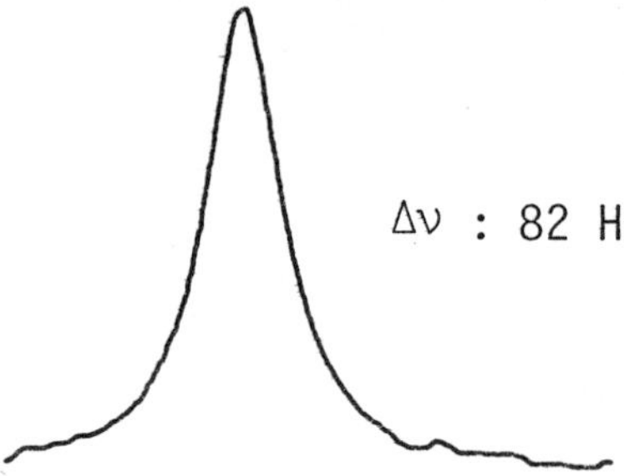

$\Delta\nu$: 82 Hz

Spectral width (Hz)	:	5 000
Number of data points	:	1 K
Pulse	:	11 μs/ 90°
Number of scans	:	100
Repetition rate (s)	:	0.1
^1H decoupled	:	-
Exp. filter (Hz)	:	10

CHEMICAL SHIFTS : reference $Cu(MeCN)_4^+BF_4^-$, MeCN

Cu(I)

ppm + 200 0 - 200

COUPLING CONSTANTS : typical values (Hz)

	^1H	^{19}F	^{31}P	^{13}C	Homo
$\lvert ^1J \rvert$ (^{63}Cu)	- to -	- to -	1109 to 1458	- to -	- to -

RELAXATION :

T_1 typical values (s) : Probably below 10^{-2}

LITTERATURE :

B. 12 p. 258 ; B 21, Vol. 10 A (1980)

O. Lutz et al. : Z. Naturforsch. 35 a, 221, (1980)

REMARKS : Only Cu (I) leads to high resolution NMR spectra. Easily detected nucleus.

ZINC Z 30

Sample	:	$Zn(NO_3)_2$	Lock	:	^2H
Solvent	:	D_2O	Bo	:	5.875
Concentration	:	2 M	Temperature	:	27°

^{67}Zn

Spin	:	5/2
Nat. abund (%)	:	4.11
Receptivity / ^{13}C	:	0.665
Gyromagnetic ratio	:	1.6737
Quad. moment	:	0.15

Resonance Freq. (MHz) referred to ^1H TMS resonance at 100MHz

6.256_7

$\Delta\nu$: 116 Hz

Spectral width (Hz)	:	6 000
Number of data points	:	4 K
Pulse	:	70 µs / 90°
Number of scans	:	2 000
Repetition rate (s)	:	0.34
^1H decoupled	:	-
Exp. filter (Hz)	:	50

CHEMICAL SHIFTS : reference $ZnClO_4$, aqueous

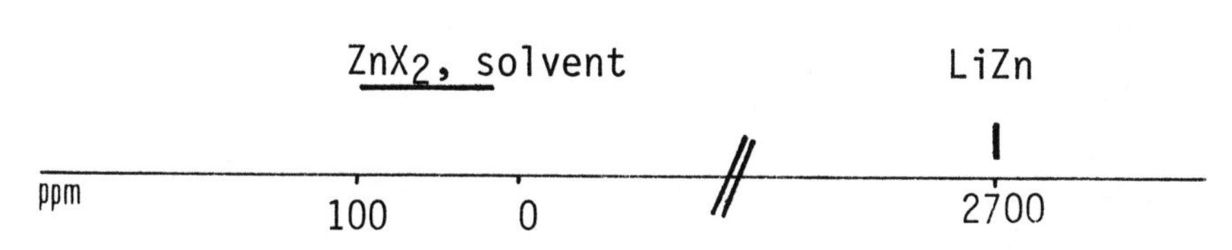

COUPLING CONSTANTS : typical values (Hz)

	^1H	^{19}F	^{31}P	^{13}C	Homo
^1J	- to -	- to -	- to -	- to -	- to -

RELAXATION :

T_1 typical values (s) : $2.\ 10^{-2}$ to 5.10^{-3}

LITTERATURE :

B.12 p. 260 ; B. 21 Vol. 10 A (1980)

O. Lutz et al. : Z. Naturforsch. 29 a 1553 (1974)

REMARKS : Linewidths are very solvent and concentration dependent

GALLIUM Z 31

Sample	:	$Ga(NO_3)_3$	Lock	:	^2H
Solvent	:	D_2O	Bo	:	2.114
Concentration	:	1.3 M/1	Temperature	:	27°

^{69}Ga

Spin	:	3/2
Nat abund (%)	:	60.4
Receptivity / ^{13}C	:	237
Gyromagnetic ratio	:	6.420
Quad. moment	:	0.178

Resonance Freq. (MHz) referred to ^1H TMS resonance at 100 MHz

24.001_4

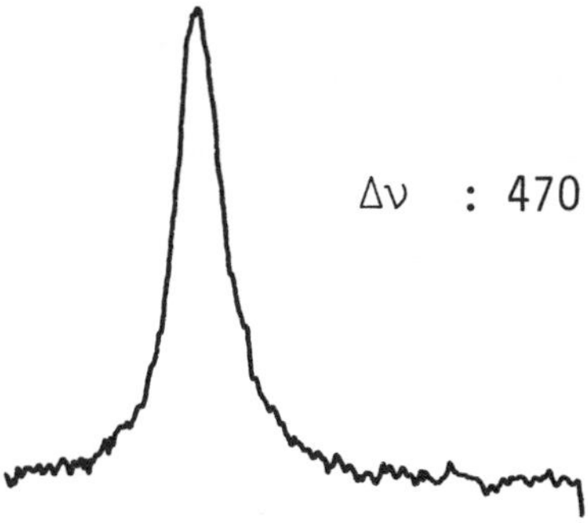

$\Delta\nu$: 470 Hz

Spectral width (Hz)	:	10 000
Number of data points	:	1 K
Pulse	:	13 µs/ 80°
Number of scans	:	1 500
Repetition rate (s)	:	0.05
^1H decoupled	:	-
Exp. filter (Hz)	:	10

^{71}Ga

Spin	:	3/2
Nat. abund (%)	:	39.6
Receptivity / ^{13}C	:	319
Gyromagnetic ratio	:	8.158
Quad. moment	:	0.112

Resonance Freq. (MHz) referred to ^1H TMS resonance at 100 MHz

30.496_7

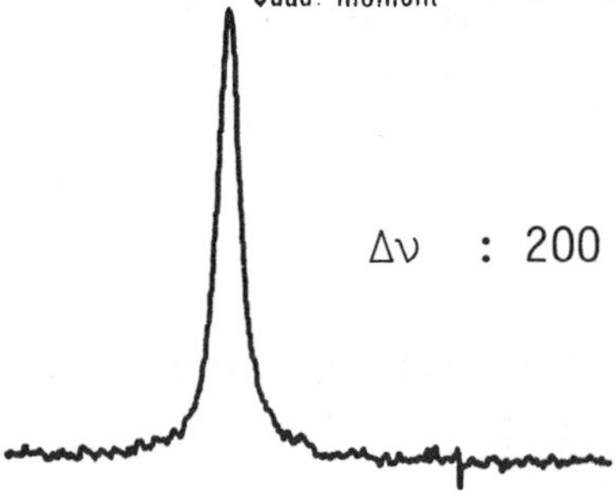

$\Delta\nu$: 200 Hz

Spectral width (Hz)	:	5 000
Number of data points	:	1 K
Pulse	:	13 µs/ 90°
Number of scans	:	100
Repetition rate (s)	:	0.1
^1H decoupled	:	-
Exp. filter (Hz)	:	10

CHEMICAL SHIFTS : reference $Ga(H_2O)_6^{3+}$

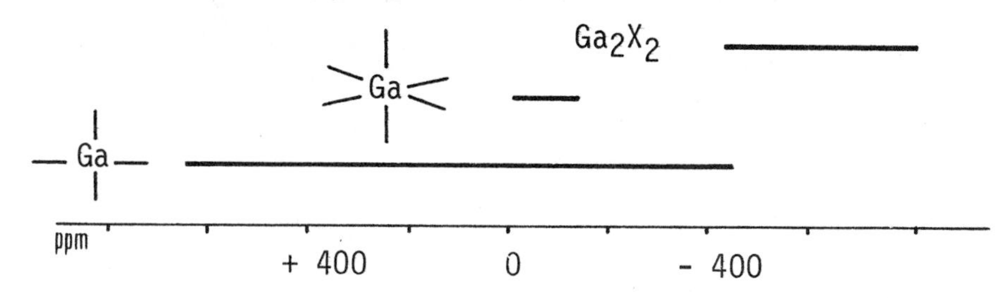

COUPLING CONSTANTS : typical values (Hz)

	^1H	^{19}F	^{31}P	^{13}C	Homo
$\vert ^1J \vert$ (^{71}Ga)	650	- to -	- to -	- to -	- to -

RELAXATION :

T_1 typical values (s) : 10^{-2} and below

LITTERATURE :

B. 12 p. 285

J.W. Akitt, (B 21) Vol. 5 A, p. 466 (1972)

REMARKS : ^{71}Ga has to be preferred (higher receptivity, sharper lines)

GERMANIUM Z 32

Sample	: $GeMe_4$	Lock	:	^2H
Solvent	: C_6D_6	Bo	:	5.875
Concentration	: 90/10 (v/v)	Temperature	:	27°

^{73}Ge

Spin	:	9/2
Nat. abund (%)	:	7.76
Receptivity / ^{13}C	:	0.617
Gyromagnetic ratio	:	-0.9331
Quad. moment	:	-0.2

Resonance Freq. (MHz) referred to ^1H TMS resonance at 100MHz

3.488_3

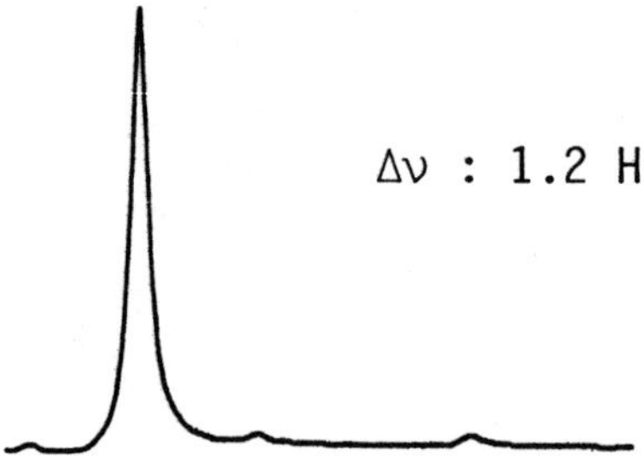

$\Delta\nu$: 1.2 Hz

Spectral width (Hz)	:	1 000
Number of data points	:	4 K
Pulse	:	60 μs / 90°
Number of scans	:	10
Repetition rate (s)	:	2.04
^1H decoupled	:	yes
Exp. filter (Hz)	:	1

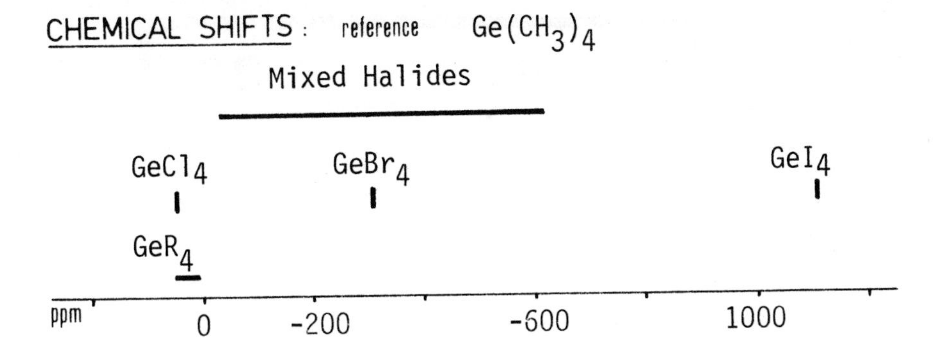

COUPLING CONSTANTS : typical values (Hz)

	^1H	^{19}F	^{31}P	^{13}C	Homo
$\vert ^1J \vert$	97.6 (GeH_4)	178.5 (GeF_4)	- to -	18.7 (GeMe4)	- to -

RELAXATION :

T_1 typical values (s) : 0.03 to 1

LITTERATURE :

B. 12 p. 340

REMARKS :

ARSENIC Z 33

Sample : $NaAsF_6$
Solvent : CH_3CN
Concentration : half sat.

Lock : C_6F_6 Cap.
Bo : 2.114
Temperature : 27°

^{75}As

Spin : 3/2
Nat. abund (%) : 100
Receptivity / ^{13}C : 143
Gyromagnetic ratio : 4.5804
Quad. moment : 0.3

Resonance Freq. (MHz) referred to ^1H TMS resonance at 100MHz

17.122_7

$\Delta\nu$: 180 Hz

Spectral width (Hz) : 10 000
Number of data points : 1 K
Pulse : 15 μs / 90°
Number of scans : 2 500
Repetition rate (s) : 0.05
^1H decoupled : -
Exp. filter (Hz) : 20

CHEMICAL SHIFTS : reference $K\ AsF_6$

AsO_4^{3-} (+ 369) AsH_4^{+} (- 291)

Organic

ppm   + 200   0   - 200

COUPLING CONSTANTS : typical values (Hz)

	^1H	^{19}F	^{31}P	^{13}C	Homo
$\lvert ^1J \rvert$	92 to 555	840 to 1048	- to -	- to -	- to -

RELAXATION :

T_1 typical values (s) : 10^{-2} to 10^{-4}

LITTERATURE :

B. 12   p. 379

G. Baliman and P.S. Pregosin, J. Magn. Reson., 26, 283, (1977)

REMARKS : Detectable only in very symmetrical environment

SELENIUM Z 34

Sample	: $SeMe_2$		Lock	: ^2H
Solvent	: C_6D_6		Bo	: 2.114
Concentration	: 98/2 (v/v)		Temperature	: 27°

^{77}Se

Spin	: 1/2
Nat. abund (%)	: 7.58
Receptivity / ^{13}C	: 2.98
Gyromagnetic ratio	: 5.1018
Quad. moment	: -

Resonance Freq. (MHz) referred to ^1H TMS resonance at 100MHz

19.071_5

$\Delta\nu$: 1 Hz

Spectral width (Hz)	: 1 000
Number of data points	: 4 K
Pulse	: 20 µs / 80°
Number of scans	: 10
Repetition rate (s)	: 2
^1H decoupled	: yes
Exp. filter (Hz)	: 0.5

CHEMICAL SHIFTS : reference SeMe$_2$ neat

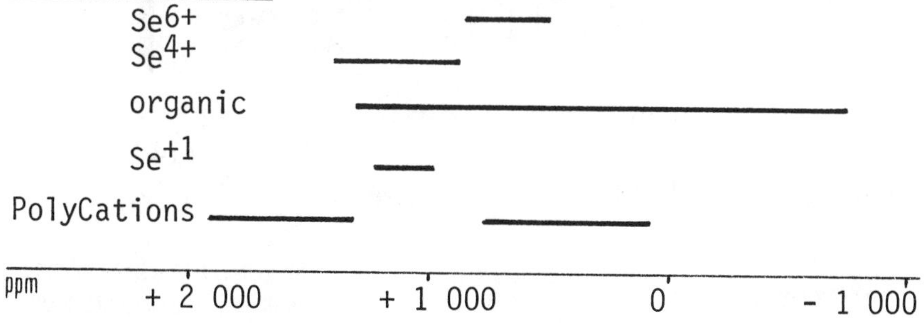

Se^{6+}

Se^{4+}

organic

Se^{+1}

PolyCations

ppm + 2 000 + 1 000 0 - 1 000

COUPLING CONSTANTS : typical values (Hz)

	^1H	^{19}F	^{31}P	^{13}C	Homo
$\lvert ^1J \rvert$	43 to 65	302 to 1584	205 to 1027	13 to 88	13 to 188

RELAXATION :

T_1 typical values (s) : 31 to 0.1

LITTERATURE :

B. 12 p. 402

P.D. Ellis et al., J. Am. Chem. Soc., 101, 5815 (1979)

REMARKS : Easy to observe. No NOE except when proton is directly bound to Se.

BROMINE Z 35

Sample	: NaBr	Lock	: ^2H
Solvent	: D_2O	Bo	: 2.114
Concentration	: 9.8 M	Temperature	: 27°

^{81}Br

Spin	: 3/2
Nat abund (%)	: 49.46
Receptivity / ^{13}C	: 277
Gyromagnetic ratio	: 7.2246
Quad. moment	: 0.28

Resonance Freq. (MHz) referred to ^1H TMS resonance at 100 MHz

27.007_0

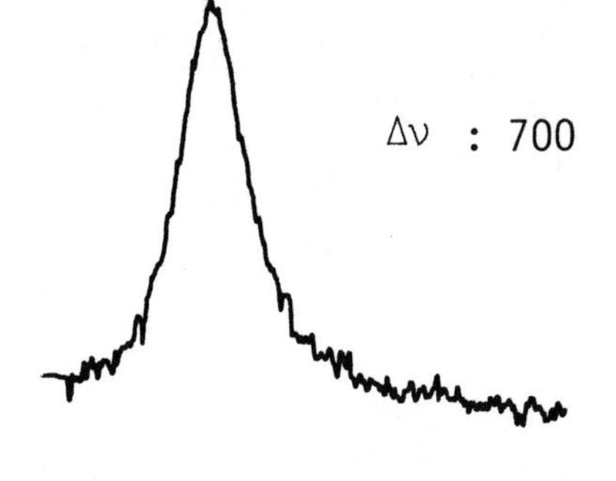

$\Delta\nu$: 700 Hz

Spectral width (Hz)	: 6 000
Number of data points	: 2 K
Pulse	: 10 μs/ 70°
Number of scans	: 4
Repetition rate (s)	: 0.17
^1H decoupled	: -
Exp. filter (Hz)	: 10

^{79}Br

Spin	: 3/2
Nat. abund (%)	: 50.54
Receptivity / ^{13}C	: 226
Gyromagnetic ratio	: 6.7023
Quad. moment	: 0.33

Resonance Freq. (MHz) referred to ^1H TMS resonance at 100 MHz

25.054_4

$\Delta\nu$: 950 Hz

Spectral width (Hz)	: 6 000
Number of data points	: 2 K
Pulse	: 10 μs/ 70°
Number of scans	: 37
Repetition rate (s)	: 0.17
^1H decoupled	: -
Exp. filter (Hz)	: 10

CHEMICAL SHIFTS : reference Br^- Na/H_2O

The knowledge of higher oxydationstates.
(BrO_3^-, BrO_4^-) may enlarge this range dowfield
(2 000 ppm or more)

←—— .... ?

ppm + 100 0 - 100 - 300

COUPLING CONSTANTS : typical values (Hz)

	^1H	^{19}F	^{31}P	^{13}C	Homo
$\lvert ^1J \rvert$ (^{81}Br)	62 (HBr)	1 697 (BrF_6^+)	315 to 380	- to -	- to -

RELAXATION :

T_1 typical values (s) :

covalent : below 10^{-7}

ionic : below $2 10^{-3}$

LITTERATURE :

B. 12 p. 421 ; B. 8 Vol. 12 (1976)

S. Forsen and B. Lindman in "Method of Biochemical Analysis, Vol. 27 (1981), John Wiley

REMARKS : Covalent Bromine is generally unobservable.
Ionic species give broad lines.

RUBIDIUM Z 37

Sample	: RbCl		Lock	: ^2H
Solvent	: D_2O		Bo	: 2.114
Concentration	: 5.6 M		Temperature	: 27°

^{85}Rb

Spin	: 5/2
Nat abund (%)	: 72.15
Receptivity / ^{13}C	: 43
Gyromagnetic ratio	: 2.5828
Quad. moment	: 0.25

Resonance Freq. (MHz) referred to ^1H TMS resonance at 100 MHz

9.655_1

Δν : 150 Hz

Spectral width (Hz)	: 6 000
Number of data points	: 4 K
Pulse	: 30 μs/ 60°
Number of scans	: 25
Repetition rate (s)	: 0.34
^1H decoupled	: -
Exp. filter (Hz)	: 10

^{87}Rb

Spin	: 3/2
Nat. abund (%)	: 27.85
Receptivity / ^{13}C	: 277
Gyromagnetic ratio	: 8.7532
Quad. moment	: 0.12

Resonance Freq (MHz) referred to ^1H TMS resonance at 100 MHz

32.721_2

Δν : 132 Hz

Spectral width (Hz)	: 6 000
Number of data points	: 8 K
Pulse	: 15 μs/ 80°
Number of scans	: 1
Repetition rate (s)	: -
^1H decoupled	: -
Exp. filter (Hz)	: 4

CHEMICAL SHIFTS : reference $RbCl/H_2O$

COUPLING CONSTANTS : typical values (Hz)

	^1H	^{19}F	^{31}P	^{13}C	Homo
^1J	- to -	- to -	- to -	- to -	- to -

RELAXATION :

T_1 typical values (s) : 0.003 and below

LITTERATURE :

B. 12 p. 129

A.I. Popov, Pure and Appl. Chem., 51, 101 (1979)

REMARKS : Easily observed in the ionic state.
But rather broad lines.

STRONTIUM Z 38

Sample	: $SrCl_2$	Lock	:	^2H
Solvent	: D_2O	Bo	:	5.875
Concentration	: 0.5 M	Temperature	:	27°

^{87}Sr

Spin	: 9/2
Nat. abund (%)	: 7.02
Receptivity / ^{13}C	: 1.07
Gyromagnetic ratio	: -1.1593
Quad. moment	: 0.36

Resonance Freq. (MHz) referred to ^1H TMS resonance at 100MHz

4.333_8

$\Delta\nu$: 131 Hz

Spectral width (Hz)	: 5 000
Number of data points	: 2 K
Pulse	: 80 μs / 90°
Number of scans	: 8 000
Repetition rate (s)	: 0.204
^1H decoupled	: -
Exp. filter (Hz)	: 50

CHEMICAL SHIFTS : reference Sr^{2+}/H_2O

$[Sr^{2+}]$, solvent

ppm + 20 0 - 20

COUPLING CONSTANTS : typical values (Hz)

	^1H	^{19}F	^{31}P	^{13}C	Homo
^1J	- to -	- to -	- to -	- to -	- to -

RELAXATION :

T_1 typical values (s) : $5\ 10^{-3}$ and below

LITTERATURE :

B. 12 p. 183

J. Banck and A. Schwenk, Z. Physik, 265, 165 (1973)

REMARKS : Its low sensitivity and its large quadrupole moment limit its study to ionic species.

YTTRIUM Z 39

Sample	: $Y(NO_3)_3$	Lock	:	^2H
Solvent	: D_2O	Bo	:	5.875
Concentration	: 2 M	Temperature	:	27°

^{89}Y

Spin	: 1/2
Nat. abund (%)	: 100
Receptivity / ^{13}C	: 0.668
Gyromagnetic ratio	: -1.3108
Quad. moment	: -

Resonance Freq. (MHz) referred to ^1H TMS resonance at 100MHz

4.900_2

*

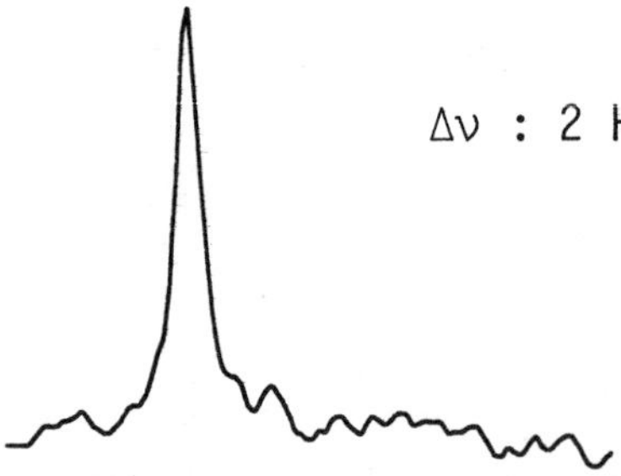

Δν : 2 Hz

Spectral width (Hz)	: 1 000
Number of data points	: 4 K
Pulse	: 30 μs / 45°
Number of scans	: 200
Repetition rate (s)	: 17
^1H decoupled	: -
Exp. filter (Hz)	: 1

CHEMICAL SHIFTS : reference $Y(ClO_4)_3$, M D_2O

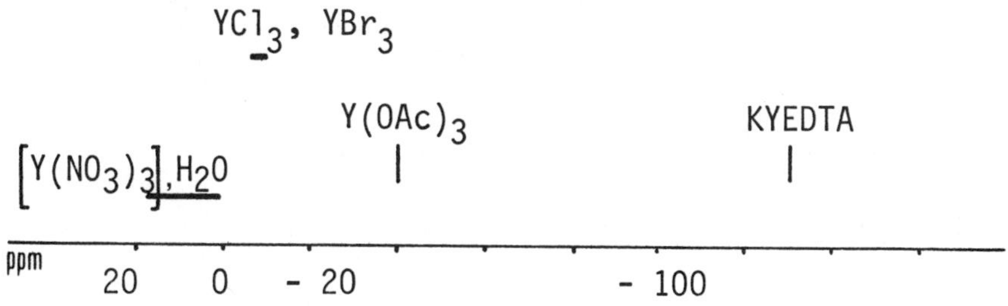

YCl_3, YBr_3

$[Y(NO_3)_3]$, H_2O — $Y(OAc)_3$ — KYEDTA

ppm 20 0 - 20 - 100

COUPLING CONSTANTS : typical values (Hz)

	^1H	^{19}F	^{31}P	^{13}C	Homo
$\|^1J\|$	- to -	- to -	- to -	12.2	- to -

RELAXATION :

T_1 typical values (s) : > 50

LITTERATURE :

B. 12 p. 199 ; B. 21 Vol. 10 A (1980)

D.G. Reid et al. : J. Magn. Res., 33, 655 (1979)

G.C. Levy et al. : J. Magn. Res., 40, 167 (1980)

REMARKS : Potential intermolecular NOE effect in Y^{3+}, H_2O solutions

ZIRCONIUM Z 40

Sample	: $(C_5H_5)_2ZrCl_2$	Lock	:	unlocked
Solvent	: CH_2Cl_2	Bo	:	5.875
Concentration	: saturated	Temperature	:	27°

^{91}Zr

Spin	: 5/2
Nat. abund (%)	: 11.23
Receptivity / ^{13}C	: 6.04
Gyromagnetic ratio	: -2.4868
Quad. moment	: (-)

Resonance Freq. (MHz) referred to ^1H TMS resonance at 100MHz

9.296_2

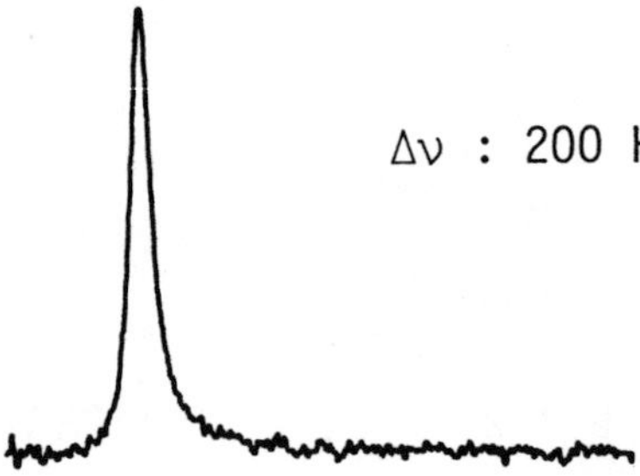

$\Delta\nu$: 200 Hz

Spectral width (Hz)	: 6 000
Number of data points	: 2 K
Pulse	: 50 μs / 90°
Number of scans	: 6 000
Repetition rate (s)	: 0.10
^1H decoupled	: -
Exp. filter (Hz)	: 10

CHEMICAL SHIFTS: reference to be defined

ppm

COUPLING CONSTANTS: typical values (Hz)

	^{1}H	^{19}F	^{31}P	^{13}C	Homo
^{1}J	- to -	- to -	- to -	- to -	- to -

RELAXATION :

T_1 typical values (s) : $< 10^{-2}$

LITTERATURE : B. 21 Vol. 10 A (1980)

M.J. McGlinchey et al., Inorg. Chim. Acta, 48, 51 (1981)

REMARKS: Zr NMR should be a valuable structural tool for inorganic chemists.

NIOBIUM Z 41

Sample	:	$KNbCl_6$	Lock	:	^2H
Solvent	:	CD_3CN	Bo	:	2.114
Concentration	:	saturated	Temperature	:	27°

^{93}Nb

Spin	:	9/2
Nat. abund (%)	:	100
Receptivity / ^{13}C	:	2740
Gyromagnetic ratio	:	6.5476
Quad. moment	:	-0.2

Resonance Freq. (MHz) referred to ^1H TMS resonance at 100MHz

24.4761_1

$\Delta\nu$: 14 Hz

Spectral width (Hz)	:	5 000
Number of data points	:	4 K
Pulse	:	16 μs / 90°
Number of scans	:	1
Repetition rate (s)	:	-
^1H decoupled	:	-
Exp. filter (Hz)	:	2

CHEMICAL SHIFTS : reference $NbCl_6$

$\delta_{NbF_6^-} = -1\,490$ ppm

Nb(V)

ppm + 400 0 - 400 - 1000

COUPLING CONSTANTS : typical values (Hz)

	^1H	^{19}F	^{31}P	^{13}C	Homo
$\lvert ^1J \rvert$	- to -	334 to 345	1050	- to -	- to -

RELAXATION :

T_1 typical values (s) : Below 10^{-1}

LITTERATURE :

B. 12 p. 208

R.G. Kidd "Nuclear Shielding of the Transition Metals" B. 21, Vol. 10 A (1980)

REMARKS : The fourth most sensitive nucleus after ^3H, ^1H and ^{19}F. Only Nb (V) has been subject to NMR investigation.

MOLYBDENUM Z 42

Sample	:	MoO_4Na_2	Lock	:	^2H
Solvent	:	D_2O	Bo	:	5.875
Concentration		2 M	Temperature	:	27°

^{95}Mo

Spin	:	5/2
Nat abund (%)	:	15.72
Receptivity / ^{13}C	:	2.88
Gyromagnetic ratio	:	1.7433
Quad. moment	:	0.12

Resonance Freq. (MHz) referred to ^1H TMS resonance at 100 MHz

6.516_9

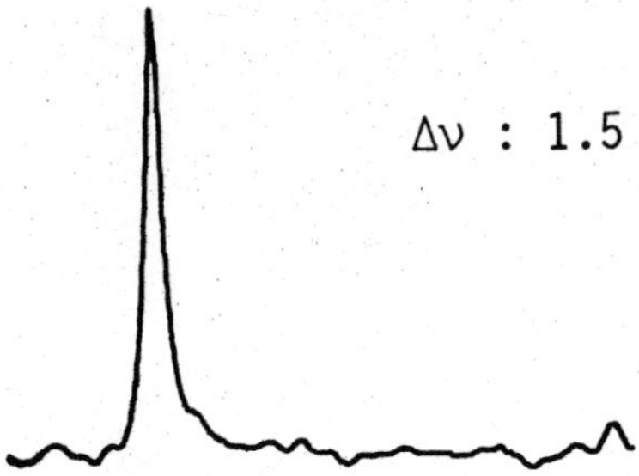

Δν : 1.5 Hz

Spectral width (Hz)	:	2 000
Number of data points	:	8 K
Pulse	:	70 μs/ 90°
Number of scans	:	1
Repetition rate (s)	:	-
^1H decoupled	:	-
Exp. filter (Hz)	:	1

^{97}Mo

Spin	:	5/2
Nat. abund (%)	:	9.46
Receptivity / ^{13}C	:	1.84
Gyromagnetic ratio	:	-1.7799
Quad. moment	:	1.1

Resonance Freq. (MHz) referred to ^1H TMS resonance at 100 MHz

6.653_6

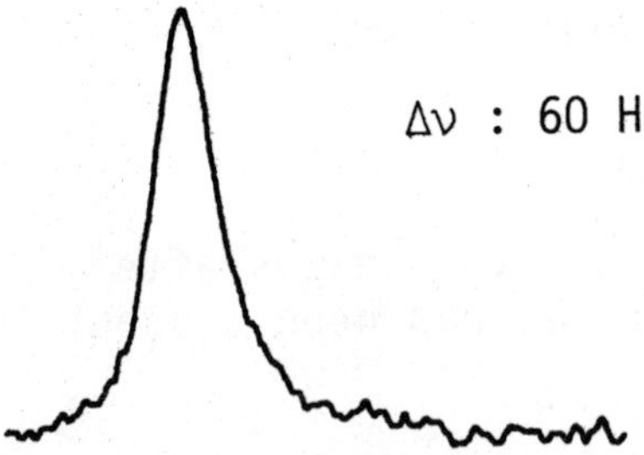

Δν : 60 Hz

Spectral width (Hz)	:	4 000
Number of data points	:	8 K
Pulse	:	60 μs/ 90°
Number of scans	:	550
Repetition rate (s)	:	1.02
^1H decoupled	:	-
Exp. filter (Hz)	:	5

CHEMICAL SHIFTS : reference MoO_4^{2-}, aqueous

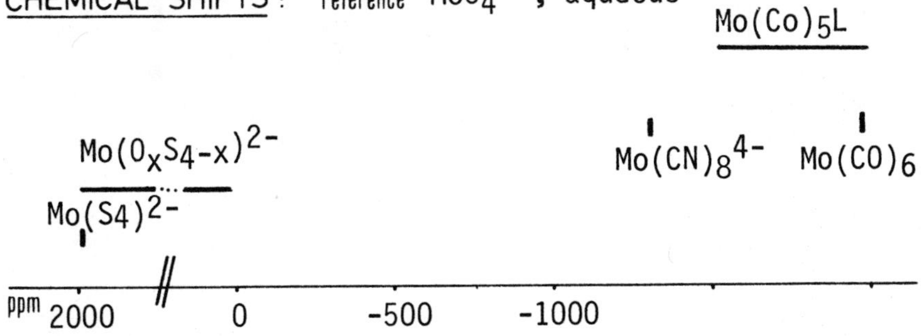

$Mo(Co)_5L$

$Mo(O_xS_{4-x})^{2-}$

$Mo(S_4)^{2-}$

$Mo(CN)_8^{4-}$ $Mo(CO)_6$

ppm 2000 // 0 -500 -1000

COUPLING CONSTANTS : typical values (Hz)

	^1H	^{19}F	^{31}P	^{13}C	Homo
$\mid ^1J \mid$ (^{95}Mo)	- to -	- to -	120 to 300	68 $Mo(CO)_6$	- to -

RELAXATION :

T_1 typical values (s) :

^{95}Mo : 0.2 to 6.10-3

^{97}Mo : 10-3 to 10^{-4}

LITTERATURE :

B. 12 p. 213 ; B.21 Vol. 10 A (1980)

J. Cotton et al., J. Organometal. Chem., 195 C17 (1980)

REMARKS : ^{95}Mo gives always sharper lines compared to ^{97}Mo

TECHNETIUM Z 43

Sample	: TcO_4NH_4	Lock	: ^2H
Solvent	: D_2O	Bo	: 2.114
Concentration	: half saturated	Temperature	: 27°

^{99}Tc

Spin	: 9/2
Nat. abund (%)	: 100
Receptivity / ^{13}C	: 2 134
Gyromagnetic ratio	: 6.0211
Quad. moment	: 0.3

Resonance Freq. (MHz) referred to ^1H TMS resonance at 100MHz

22.508_3

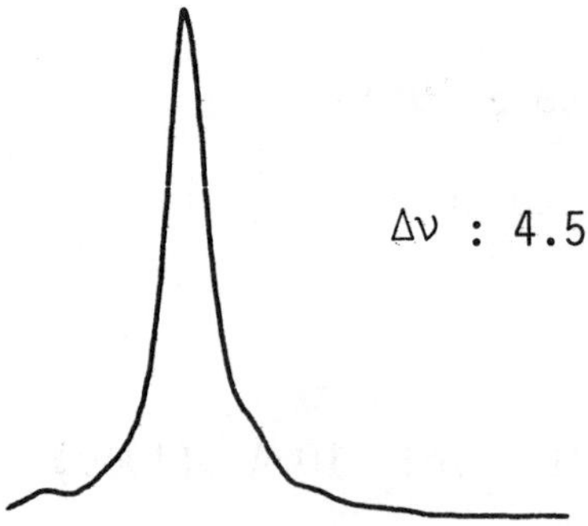

$\Delta\nu$: 4.5 Hz

Spectral width (Hz)	: 1 000
Number of data points	: 4 K
Pulse	: 15 μs / 70°
Number of scans	: 1
Repetition rate (s)	: -
^1H decoupled	: -
Exp. filter (Hz)	: 1

CHEMICAL SHIFTS : reference TcO_4^-

Only TcO_4 has been observed

ppm

COUPLING CONSTANTS : typical values (Hz)

	^1H	^{19}F	^{31}P	^{13}C	Homo
^1J	- to -	- to -	- to -	- to -	- to -

RELAXATION :

T_1 typical values (s) : $2\ 10^{-1}$ and below

LITTERATURE :

B. 12 p. 222

R.G. Kidd "Nuclear Shielding of the Transition Metals"
B. 21, Vol. 10 A (1980)

REMARKS : Only ^{99}Tc (half life : $5\ 10^{+5}$ years) is of interest for the NMR spectroscopist. The fifth most sensitive nucleus after ^3H, ^1H, ^{19}F and ^{93}Nb.

RUTHENIUM Z 44

Sample	:	RuO_4	Lock	:	unlocked
Solvent	:	CCl_4	Bo	:	5.875
Concentration	:	1.2 M	Temperature	:	27°

^{99}Ru

Spin	:	5/2
Nat abund (%)	:	12.72
Receptivity / ^{13}C	:	0.83
Gyromagnetic ratio	:	- 1.2343
Quad. moment	:	0.076

Resonance Freq. (MHz) referred to ^1H TMS resonance at 100 MHz

4.614_0

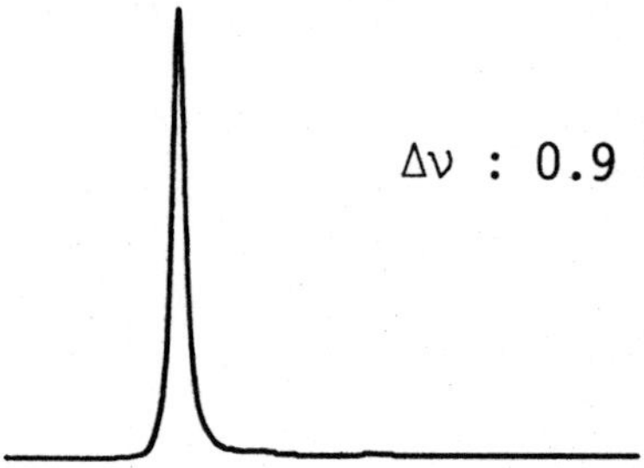

Δν : 0.9 Hz

Spectral width (Hz)	:	1 000
Number of data points	:	8 K
Pulse	:	50 μs/80°
Number of scans	:	1 000
Repetition rate (s)	:	1
^1H decoupled	:	-
Exp. filter (Hz)	:	0.2

^{101}Ru

Spin	:	5/2
Nat. abund (%)	:	17.07
Receptivity / ^{13}C	:	1.56
Gyromagnetic ratio	:	- 1.3834
Quad. moment	:	0.44

Resonance Freq. (MHz) referred to ^1H TMS resonance at 100 MHz

5.171_3

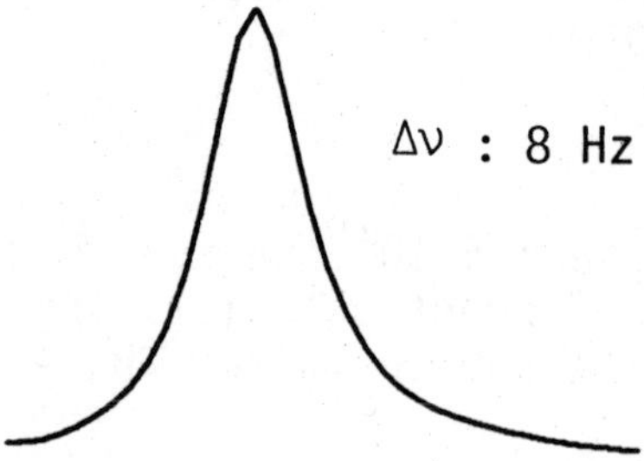

Δν : 8 Hz

Spectral width (Hz)	:	6 000
Number of data points	:	8 K
Pulse	:	65 μs/ 90°
Number of scans	:	1 200
Repetition rate (s)	:	1
^1H decoupled	:	-
Exp. filter (Hz)	:	1

CHEMICAL SHIFTS : reference RuO_4 - M / CCl_4

$Ru(biPy)_3^{2+}$ | $Ru(CN)_6^{4-}$ | $Ru_3(CO)_{12}$

ppm + 3000 0 - 3000

COUPLING CONSTANTS : typical values (Hz)

	^1H	^{19}F	^{31}P	^{13}C	Homo
$\lvert ^1J \rvert$	to	- to -	to	44.1 $Ru(CN)_6^{4-}$	- to -

RELAXATION :

T_1 typical values (s) :

^{99}Ru : 0.3 and below

^{101}Ru : 3.10^{-2} and below

LITTERATURE :

S. Buettgenbach and al : Z.Phys. 280 A, 217 (1977)

C. Brevard, P. Granger : to be published

REMARKS : ^{99}Ru gives sharp lines, easily detectable

$^1J_{Ru-^{17}O}$ = 23.4 Hz in RuO_4

RHODIUM Z 45

Sample	: $RhCl_6^{3-}$	Lock	:	^2H
Solvent	: D_2O/HCl	Bo	:	5.875
Concentration	: 0.7 M	Temperature	:	35°

^{103}Rh

Spin	:	1/2
Nat. abund (%)	:	100
Receptivity / ^{13}C	:	0.177
Gyromagnetic ratio	:	-0.8520
Quad. moment	:	-

Resonance Freq. (MHz) referred to ^1H TMS resonance at 100MHz

3.185_2

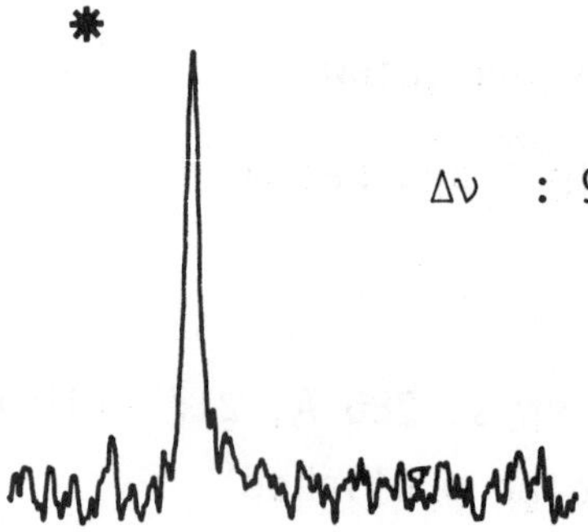

*

Δν : 9 Hz

Spectral width (Hz)	:	8 000
Number of data points	:	16 K
Pulse	:	60 μs / 40°
Number of scans	:	24 000
Repetition rate (s)	:	1
^1H decoupled	:	-
Exp. filter (Hz)	:	2

CHEMICAL SHIFTS : reference Rh metal (Ξ = 3.16 MHz)

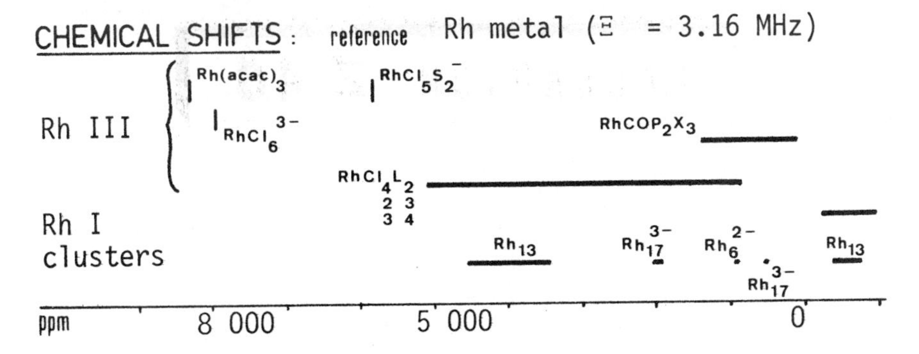

COUPLING CONSTANTS : typical values (Hz)

	^1H	^{19}F	^{31}P	^{13}C	Homo
$\lvert {}^1J \rvert$	15 to 30	55 (RhFCOPØ3)	80 to 150	33 to 80	- to -

RELAXATION :

T_1 typical values (s) : > 50

LITTERATURE :

B. 12 p. 244

B. 21 vol. 10 A (1980)

Schwenk et al. : J. Magn. Res., 41 (2), 354 (1980)
O. Gansow et al. : J. Am. Chem. Soc., 102 (7), 2449 (1980)

REMARKS : Chemical shifts are temperature dependent (ca. 2 ppm/°C)

PALLADIUM Z 46

Sample	:	K_2PdCl_6	Lock	:	unlocked
Solvent	:	D_2O	Bo	:	4.700
Concentration	:	Sat.	Temperature	:	27^c

^{105}Pd

Spin	:	5/2
Nat. abund (%)	:	22.23
Receptivity / ^{13}C	:	1.41
Gyromagnetic ratio	:	- 0.756
Quad. moment	:	+ 0.8

Resonance Freq. (MHz) referred to ^1H TMS resonance at 100MHz

4.576_1

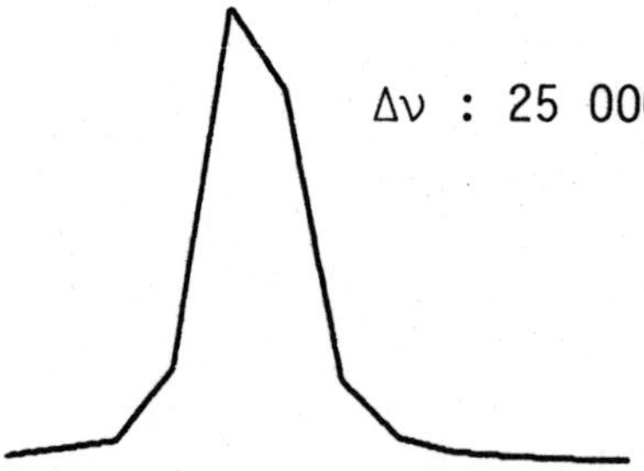

$\Delta\nu$: 25 000 Hz

Spectral width (Hz)	:	5.10^6 Hz
Number of data points	:	1 K
Pulse	:	10 μs / 90°
Number of scans	:	10.000
Repetition rate (s)	:	0.0001
^1H decoupled	:	-
Exp. filter (Hz)	:	1 000

CHEMICAL SHIFTS : reference

Only one resonance known (K_2PdCl_6)

ppm

COUPLING CONSTANTS : typical values (Hz)

	^1H	^{19}F	^{31}P	^{13}C	Homo
^1J	- to -	- to -	- to -	- to -	- to -

RELAXATION :

T_1 typical values (s) : below 10^{-5}

LITTERATURE :

J.A. Seitchich, A. Gossard, V. Jaccarino
Phys. Rev. 136 A, 1119 (1964)

REMARKS :

^{105}Pd will be of little use in high resolution NMR because of very broad lines

SILVER Z 47

Sample	: $AgNO_3$	Lock	: ^2H
Solvent	: D_2O	Bo	: 5.875
Concentration	: 2 M	Temperature	: 27°

^{107}Ag

Spin	: 1/2
Nat abund (%)	: 51.82
Receptivity / ^{13}C	: 0.195
Gyromagnetic ratio	: -1.0828
Quad. moment	: -

Resonance Freq. (MHz) referred to ^1H TMS resonance at 100 MHz

4.047_8

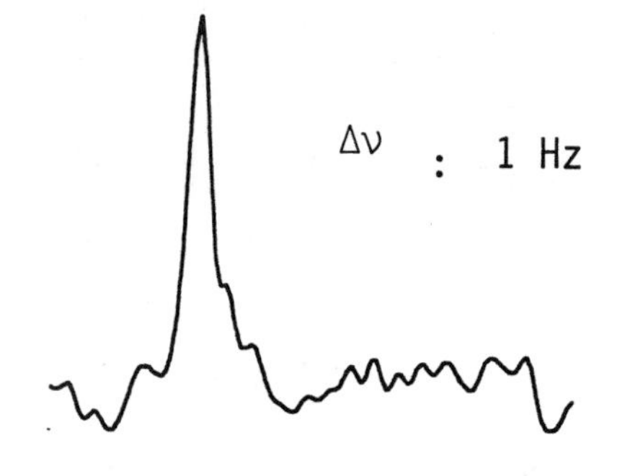

$\Delta\nu$: 1 Hz

Spectral width (Hz)	: 2 000
Number of data points	: 2 K
Pulse	: 50 μs/ 40°
Number of scans	: 33
Repetition rate (s)	: 20.5
^1H decoupled	: -
Exp. filter (Hz)	: 1

^{109}Ag

Spin	: 1/2
Nat. abund (%)	: 48.18
Receptivity / ^{13}C	: 0.276
Gyromagnetic ratio	: -1.2448
Quad. moment	: -

Resonance Freq. (MHz) referred to ^1H TMS resonance at 100 MHz

4.653_5

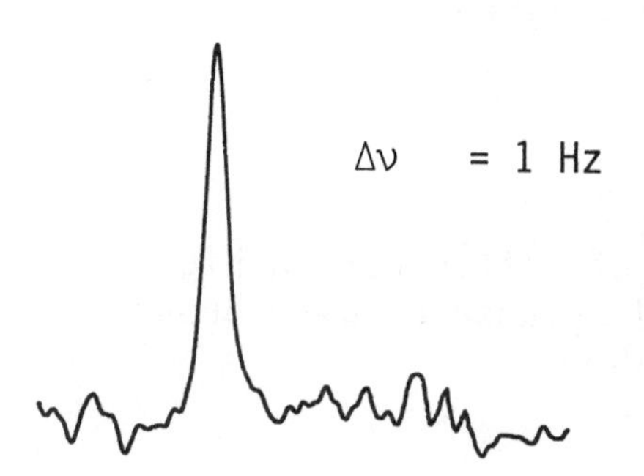

$\Delta\nu$ = 1 Hz

Spectral width (Hz)	: 1 000
Number of data points	: 2 K
Pulse	: 50 μs/ 40°
Number of scans	: 25
Repetition rate (s)	: 21
^1H decoupled	: -
Exp. filter (Hz)	: 1

CHEMICAL SHIFTS : reference Ag^+, aqueous

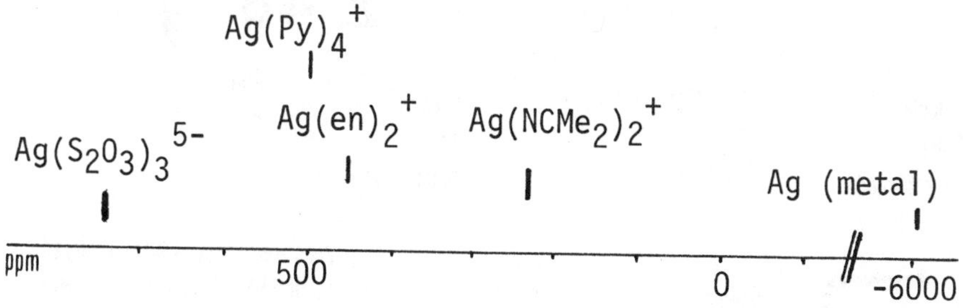

$Ag(Py)_4^+$

$Ag(S_2O_3)_3^{5-}$ $Ag(en)_2^+$ $Ag(NCMe_2)_2^+$ Ag (metal)

ppm 500 0 -6000

COUPLING CONSTANTS : typical values (Hz)

	^1H	^{19}F	^{31}P	^{13}C	Homo
$\|^1J\|$ (^{107}Ag)	- to -	- to -	220 to 760	- to -	- to -

RELAXATION :

T_1 typical values (s) : 60 to 950

LITTERATURE :

B.12 p. 258 ; B. 21 Vol. 10 A (1980)

A.I. Popov et al., J. Magn. Res., 36 351 (1979)
Schwenk et al., Z. Naturforsch., A 35 319 (1980)

REMARKS : Very long relaxation times

CADMIUM Z 48

Sample	: $Cd(CH_3COO)_2$	Lock	:	^2H
Solvent	: D_2O	Bo	:	2.114
Concentration	: 2.1 M	Temperature	:	27°

^{113}Cd

Spin	: 1/2
Nat abund (%)	: 12.26
Receptivity / ^{13}C	: 7.6
Gyromagnetic ratio	: -5.9328
Quad. moment	: -

Resonance Freq. (MHz) referred to ^1H TMS resonance at 100 MHz

22.178_1

Δν : 1.5 Hz

Spectral width (Hz)	: 500
Number of data points	: 16 K
Pulse	: 15 μs/ 70°
Number of scans	: 4
Repetition rate (s)	: 16.4
^1H decoupled	: -
Exp. filter (Hz)	: 0,5

^{111}Cd

Spin	: 1/2
Nat. abund (%)	: 12.75
Receptivity / ^{13}C	: 6.93
Gyromagnetic ratio	: -5.6714
Quad. moment	: -

Resonance Freq. (MHz) referred to ^1H TMS resonance at 100 MHz

21.200_8

Δν : 1.5 Hz

Spectral width (Hz)	: 500
Number of data points	: 16 K
Pulse	: 15 μs/ 70°
Number of scans	: 4
Repetition rate (s)	: 16.4
^1H decoupled	: -
Exp. filter (Hz)	: 0.5

CHEMICAL SHIFTS : reference $Cd(ClO_4)_2$ / H_2O

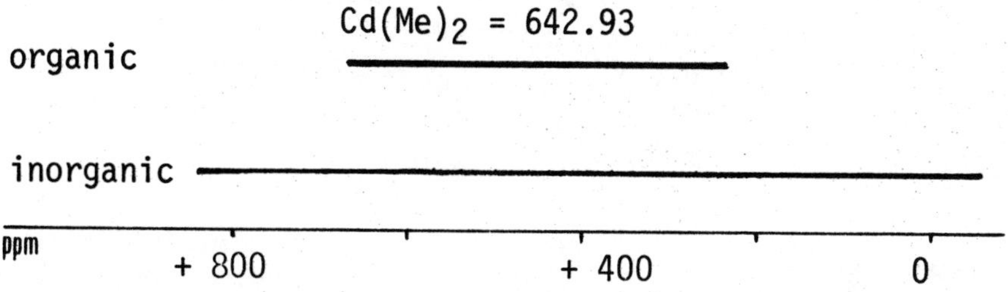

organic — $Cd(Me)_2$ = 642.93

inorganic

ppm: + 800, + 400, 0

COUPLING CONSTANTS : typical values (Hz)

	^1H	^{19}F	^{31}P	^{13}C	Homo
$\lvert ^1J \rvert$ (^{113}Cd)	- to -	- to -	1 200 to 1 710	- 500	- to -

RELAXATION :

T_1 typical values (s) : 65 to 0.1 or below

LITTERATURE :

B. 12 p. 261

S. Forsen and B. Lindman in "Methodes of Biochemical Analysis", Vol. 27 (1981), John Wiley

R. Kidd "Nuclear Shielding of the Transition Metals" B. 21 - Vol. A (1980)

REMARKS :

^{113}Cd is generally the preferred isotope but both may be used. Potential negative NOE.

INDIUM Z 49

Sample	:	$In(NO_3)_3$	Lock	:	^2H
Solvent	:	D_2O/HNO_3	Bo	:	5.875
Concentration	:	0.05 M	Temperature	:	27°

^{113}In

Spin	:	9/2
Nat abund (%)	:	4.28
Receptivity / ^{13}C	:	83.8
Gyromagnetic ratio	:	5.8493
Quad. moment	:	1.14

Resonance Freq. (MHz) referred to ^1H TMS resonance at 100 MHz

21.865_8

Δν : 1 200 Hz

Spectral width (Hz)	:	31.000
Number of data points	:	4
Pulse	:	40 μs/ 90°
Number of scans	:	10.000
Repetition rate (s)	:	0.06
^1H decoupled	:	-
Exp. filter (Hz)	:	50

^{115}In

Spin	:	9/2
Nat. abund (%)	:	95.72
Receptivity / ^{13}C	:	$1.89\ 10^3$
Gyromagnetic ratio	:	5.8618
Quad. moment	:	+ 0.83

Resonance Freq. (MHz) referred to ^1H TMS resonance at 100 MHz

21.912_6

Δν : 1 200 Hz

Spectral width (Hz)	:	30.000
Number of data points	:	4
Pulse	:	40 μs/ 90°
Number of scans	:	10.000
Repetition rate (s)	:	0.06
^1H decoupled	:	-
Exp. filter (Hz)	:	50

CHEMICAL SHIFTS : reference $\{In(H_2O)_6\}^{3+}$

$InCl_4^-$ $InBr_4^-$ InI_4^-

ppm 200 0 -200

COUPLING CONSTANTS : typical values (Hz)

	^1H	^{19}F	^{31}P	^{13}C	Homo
$\|^1J\|$ (^{115}In)	968 (InH_4^-) calculated	- to -	- to -	- to -	- to -

RELAXATION :

T_1 typical values (s) : 10^{-4}

LITTERATURE :

B 12 p. 286

REMARKS :

TIN Z 50

Sample	: $SnMe_4$	Lock	: ^2H
Solvent	: C_6D_6	Bo	: 2.114
Concentration	: 98/2 (v/v)	Temperature	: 27°

^{117}Sn

Spin	: 1/2
Nat abund (%)	: 7.61
Receptivity / ^{13}C	: 19.54
Gyromagnetic ratio	: -9.5319
Quad. moment	: -

Resonance Freq. (MHz) referred to ^1H TMS resonance at 100 MHz

35.632_2

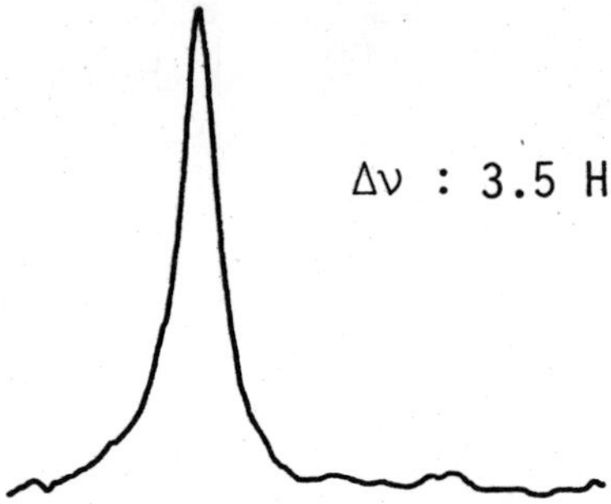

Δν : 3.5 Hz

Spectral width (Hz)	: 1 000
Number of data points	: 4 K
Pulse	: 16 μs/ 90°
Number of scans	: 4
Repetition rate (s)	: 2
^1H decoupled	: yes
Exp. filter (Hz)	: 0.5

^{119}Sn

Spin	: 1/2
Nat. abund (%)	: 8.58
Receptivity / ^{13}C	: 25.2
Gyromagnetic ratio	: -9.9756
Quad. moment	: -

Resonance Freq. (MHz) referred to ^1H TMS resonance at 100 MHz

37.290_6

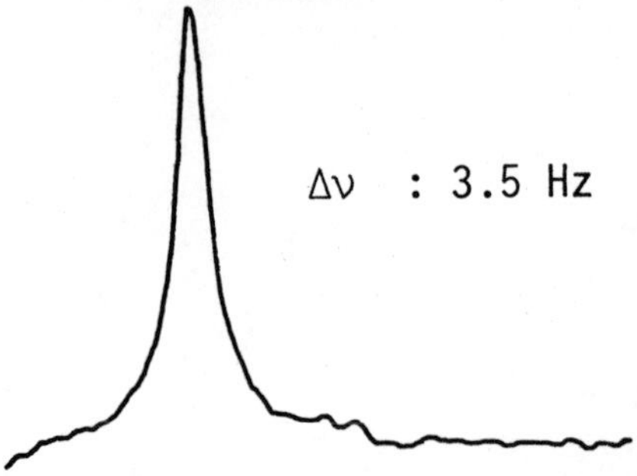

Δν : 3.5 Hz

Spectral width (Hz)	: 1 000
Number of data points	: 4 K
Pulse	: 16 μs/ 90°
Number of scans	: 3
Repetition rate (s)	: 2
^1H decoupled	: yes
Exp. filter (Hz)	: 0.5

CHEMICAL SHIFTS : reference $Sn(Me_4)$ neat

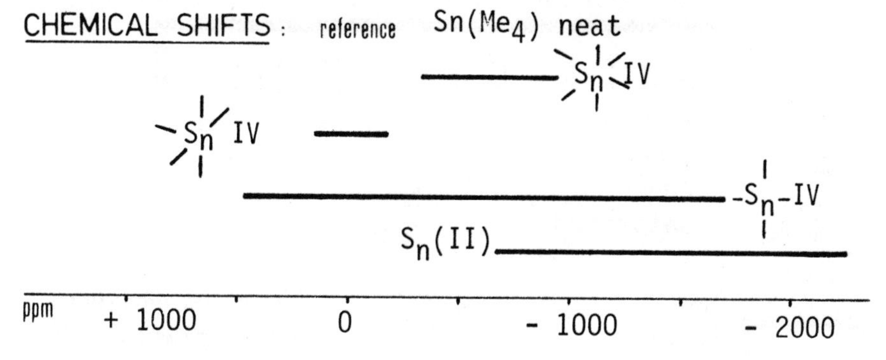

COUPLING CONSTANTS : typical values (Hz)

	^1H	^{19}F	^{31}P	^{13}C	Homo
$\vert ^1J \vert$ (^{119}Sn)	109 to 2450	128 to 2298	50 to 2383	1168 to 220	5200 to 4462

RELAXATION :

T_1 typical values (s) : 2 to 10^{-2}

LITTERATURE :

B. 12 p. 342

P.J. Smith and A.P. Tupcianskas (B. 21), Vol. 8, p. 292 (1978)

V.S. Petrosyan, B.10, Vol. 11, 115, (1977)

REMARKS : Very easy to observe. ^{119}Sn is generally preferred. Negative NOE.

ANTIMONY Z 51

Sample	:	$KSbCl_6$	Lock	:	^{2}H
Solvent	:	CD_3CN	Bo	:	2.114
Concentration	:	saturated	Temperature	:	27°

^{121}Sb

Spin	:	5/2
Nat abund (%)	:	57.25
Receptivity / ^{13}C	:	520
Gyromagnetic ratio	:	6.4016
Quad. moment	:	-0.53

Resonance Freq. (MHz) referred to ^{1}H TMS resonance at 100 MHz

23.930_6

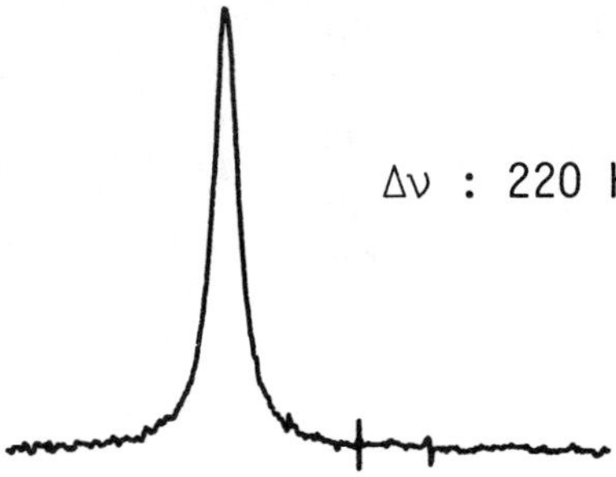

Δν : 220 Hz

Spectral width (Hz)	:	6 000
Number of data points	:	4 K
Pulse	:	15 μs/90°
Number of scans	:	50
Repetition rate (s)	:	0.34
^{1}H decoupled	:	-
Exp. filter (Hz)	:	5

^{123}Sb

Spin	:	7/2
Nat. abund (%)	:	42.75
Receptivity / ^{13}C	:	111
Gyromagnetic ratio	:	3.4668
Quad. moment	:	-0.68

Resonance Freq (MHz) referred to ^{1}H TMS resonance at 100 MHz

12.958_9

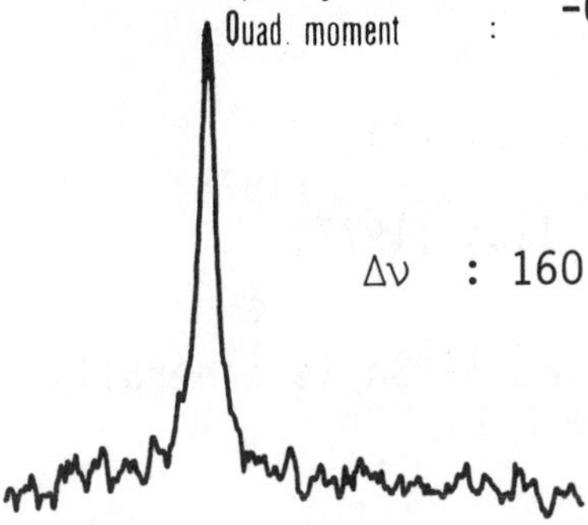

Δν : 160 Hz

Spectral width (Hz)	:	6 000
Number of data points	:	1 K
Pulse	:	25 μs/ 8Q°
Number of scans	:	1 000
Repetition rate (s)	:	0.084
^{1}H decoupled	:	-
Exp. filter (Hz)	:	20

CHEMICAL SHIFTS : reference $Et_4N^+SbCl_6^-$

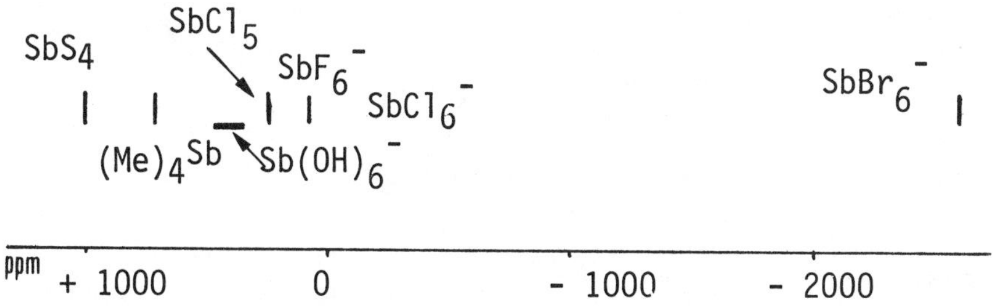

SbS_4 $SbCl_5$ SbF_6^- $SbCl_6^-$ $SbBr_6^-$

$(Me)_4Sb$ $Sb(OH)_6^-$

ppm + 1000 0 - 1000 - 2000

COUPLING CONSTANTS : typical values (Hz)

	^1H	^{19}F	^{31}P	^{13}C	Homo
$\|^1J\|$ (^{121}Sb)	- to -	1840 to 1950	- to -	- to -	- to -

RELAXATION :

T_1 typical values (s) : below 10^{-4}

LITTERATURE :

B. 12 p. 380

R.G. Kidd et al. : J. Inorg. Nucl. Chem., 37, 661 (1975)

REMARKS : ^{121}Sb is the preferred isotope. Quadrupolar relaxation prevents observation of non symmetrical compounds.

TELLURIUM Z 52

Sample	:	$(CH_3)_2Te$	Lock	:	^2H
Solvent	:	C_6D_6	Bo	:	2.114
Concentration	:	98/2 (v/v)	Temperature	:	27°

^{125}Te

Spin	:	1/2
Nat abund (%)	:	7
Receptivity / ^{13}C	:	12.5
Gyromagnetic ratio	:	-8.4398
Quad. moment	:	-

Resonance Freq. (MHz) referred to ^1H TMS resonance at 100 MHz

31.549_7

Δν : 1.5 Hz

Spectral width (Hz)	:	400
Number of data points	:	8 K
Pulse	:	12 μs/90°
Number of scans	:	1
Repetition rate (s)	:	-
^1H decoupled	:	yes
Exp. filter (Hz)	:	0.2 Hz

^{123}Te

Spin	:	1/2
Nat. abund (%)	:	0.89
Receptivity / ^{13}C	:	0.89
Gyromagnetic ratio	:	-7.0006
Quad. moment	:	-

Resonance Freq. (MHz) referred to ^1H TMS resonance at 100 MHz

26.169_7

Δν : 3 Hz

Spectral width (Hz)	:	400
Number of data points	:	8 K
Pulse	:	15 μs/ 90°
Number of scans	:	32
Repetition rate (s)	:	10.25
^1H decoupled	:	yes
Exp. filter (Hz)	:	0.5 Hz

CHEMICAL SHIFTS : reference $Te(CH_3)_2$ neat

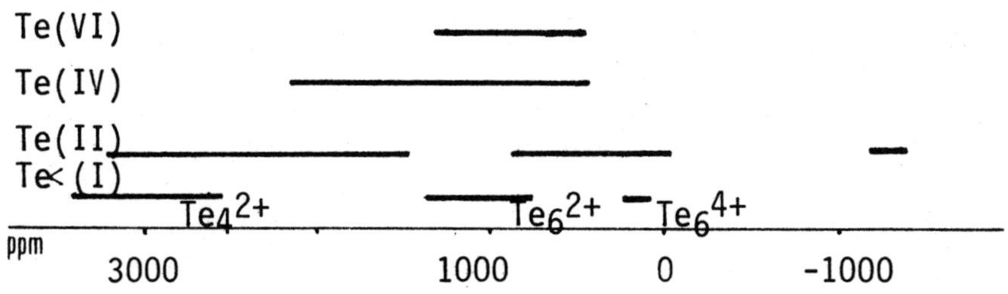

Te(VI)
Te(IV)
Te(II)
Te< (I)
Te_4^{2+} Te_6^{2+} Te_6^{4+}
ppm
3000 1000 0 -1000

COUPLING CONSTANTS : typical values (Hz)

	^1H	^{19}F	^{31}P	^{13}C	Homo
$\|^1J\|$ (^{125}Te)	59	3000 to 4000	1300 to 2000	135 to 330	84 to 265

RELAXATION :

T_1 typical values (s) : 2.5 to 0.5

LITTERATURE :

B. 12 p. 412

Most of the references on Te may be found in :
P. Granger and S. Chapelle : J. Magn. Reson., 39 329 (1980)

REMARKS : Large temperature effect on chemical shifts (0.16 ppm/°K).NoNOE. Very sensitive to structural environnment

IODINE Z 53

Sample	: KI		Lock	: C_6F_6 cap.
Solvent	: H_2O		Bo	: 2.114
Concentration	: 5.67 M		Temperature	: 27°

^{127}I

Spin	: 5/2
Nat. abund (%)	: 100
Receptivity / ^{13}C	: 530
Gyromagnetic ratio	: 5.3525
Quad. moment	: -0.79

Resonance Freq. (MHz) referred to ^1H TMS resonance at 100MHz

20.008_6

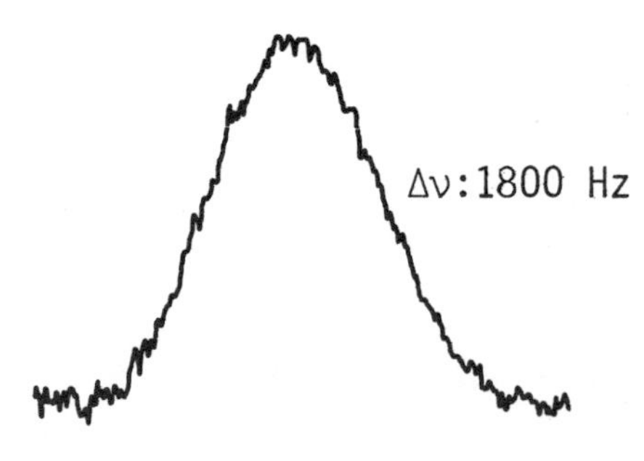

$\Delta\nu$:1800 Hz

Spectral width (Hz)	: 10 000
Number of data points	: 1 K
Pulse	: 15 μs / 80°
Number of scans	: 1 000
Repetition rate (s)	: -
^1H decoupled	: -
Exp. filter (Hz)	: 20

CHEMICAL SHIFTS : reference I^- (NaI/H_2O)

COUPLING CONSTANTS : typical values (Hz)

	^1H	^{19}F	^{31}P	^{13}C	Homo
$\lvert ^1J \rvert$	- to -	2100 to 2730	to	< 60	to

RELAXATION :

T_1 typical values (s) :

covalent : below 10^{-7}

ionic : below 5 10^{-4}

LITTERATURE :

B. 12 p. 421

B. Lindman and S. Forsen : (B8) Vol. 12, "Chlorine, Bromine and Iodine NMR", (1976)

REMARKS : High resolution spectra can only be obtained in very symmetric surrounding.

XENON Z 54

Sample	Xe Gas	Lock	unlocked
Solvent	-	Bo	2.114
Concentration	14 Bars	Temperature	27°

^{129}Xe

Spin	1/2
Nat abund (%)	26.44
Receptivity / ^{13}C	31.8
Gyromagnetic ratio	- 7.4003
Quad. moment	-

Resonance Freq. (MHz) referred to ^{1}H TMS resonance at 100MHz

27.658_1

$\Delta\nu$: 1.7 Hz

Spectral width (Hz)	1 000
Number of data points	8 K
Pulse	15 μs/ 90°
Number of scans	1
Repetition rate (s)	-
^{1}H decoupled	-
Exp. filter (Hz)	1

^{131}Xe

Spin	3/2
Nat. abund (%)	21.18
Receptivity / ^{13}C	3.31
Gyromagnetic ratio	2.1939
Quad. moment	- 0.12

Resonance Freq. (MHz) referred to ^{1}H TMS resonance at 100MHz

8.199_5

$\Delta\nu$: 5 Hz

Spectral width (Hz)	3 000
Number of data points	4 K
Pulse	30 μs/ 70°
Number of scans	1 500
Repetition rate (s)	0.7
^{1}H decoupled	-
Exp. filter (Hz)	2

CHEMICAL SHIFTS : reference $XeOF_4$ neat

+ 8
+ 6
+ 4
+ 2
0
ppm + 1000 0 - 1000 - 5300

COUPLING CONSTANTS : typical values (Hz)

	^1H	^{19}F	^{31}P	^{13}C	Homo
$\lvert ^1J \rvert$ ^{129}Xe	- to -	90 to 7 200	- to -	- to -	- to -

RELAXATION :

T_1 typical values (s) : 2 600 to 50 (gas/phase)

0.8 to 0.25 (Solutions of Xe compds)

LITTERATURE :

B. 12 p. 439

G.J. Schrobilgen, S.H. Holloway, P. Granger and C. Brevard : Inorg. Chem., 17, 980 (2978)

REMARKS : ^{129}Xe is easy to observe, but large sweep widths are often required due to J_{Xe-F} values

CESIUM Z55

Sample	: $CsNO_3$		Lock	: ^2H ext.
Solvent	: D_2O		Bo	: 2.114
Concentration	: saturated		Temperature	: 27°

^{133}Cs

Spin	: 7/2
Nat. abund (%)	: 100
Receptivity / ^{13}C	: 269
Gyromagnetic ratio	: 3.5087
Quad. moment	: $-3\ 10^{-3}$

Resonance Freq. (MHz) referred to ^1H TMS resonance at 100MHz

13.116_2

$\Delta\nu$: 0.6 Hz

Spectral width (Hz)	: 600
Number of data points	: 4 K
Pulse	: 10 μs / 20°
Number of scans	: 1
Repetition rate (s)	: -
^1H decoupled	: -
Exp. filter (Hz)	: 0

CHEMICAL SHIFTS : reference Cs Br (0.5 m) H_2O

Cs^- ▬

Cs^+

ppm + 200 0 - 200

COUPLING CONSTANTS : typical values (Hz)

	^{1}H	^{19}F	^{31}P	^{13}C	Homo
^{1}J	to	to	to	to	to

no coupling known

RELAXATION :

T_1 typical values (s) : 25 and below

LITTERATURE :

B. 12 p. 129

A.I. Popov, Pure and Appl. Chem., 51, 101 (1979)

REMARKS : Very easily observed. One of the smallest quadrupole moment.

BARYUM Z 56

Sample	: $BaCl_2$	Lock	: ^2H
Solvent	: D_2O	Bo	: 5.875
Concentration	: 0.5 M	Temperature	: 27°

^{135}Ba

Spin	: 3/2
Nat abund (%)	: 6.59
Receptivity / ^{13}C	: 1.83
Gyromagnetic ratio	: 2.6575
Quad. moment	: 0.18

Resonance Freq. (MHz) referred to ^1H TMS resonance at 100 MHz

9.934_4

Δν : 780Hz

Spectral width (Hz)	: 25 000
Number of data points	: 1 K
Pulse	: 60 μs/ 90°
Number of scans	: 80 000
Repetition rate (s)	: 0.02
^1H decoupled	: -
Exp. filter (Hz)	: 100

^{137}Ba

Spin	: 3/2
Nat. abund (%)	: 11.32
Receptivity / ^{13}C	: 4.41
Gyromagnetic ratio	: 2.9728
Quad. moment	: 0.28

Resonance Freq. (MHz) referred to ^1H TMS resonance at 100 MHz

11.112_9

Δν : 2000 Hz

Spectral width (Hz)	: 25 000
Number of data points	: 1 K
Pulse	: 40 μs/ 90°
Number of scans	: 55 000
Repetition rate (s)	: 0.02
^1H decoupled	: -
Exp. filter (Hz)	: 100

CHEMICAL SHIFTS : reference $BaCl_2$, aqueous

Ba^{2+}, solvent

ppm +5 0 -5

COUPLING CONSTANTS : typical values (Hz)

	^{1}H	^{19}F	^{31}P	^{13}C	Homo
^{1}J	- to -	- to -	- to -	- to -	- to -

RELAXATION :

T_1 typical values (s) : ^{137}Ba : 10^{-4} and below

LITTERATURE :

B. 12 p. 183

O. Lutz et al. : Z. Physik A 288, 11 (1978)

REMARKS : Most of the available data concern T_1 measurements.

LANTHANUM Z 57

Sample	: $LaCl_3$		Lock	: 2H
Solvent	: D_2O		Bo	: 5.875
Concentration	: 0.05 M		Temperature	: 27°

^{138}La

Spin	: 5
Nat abund (%)	: 0.09
Receptivity / ^{13}C	: 0.43
Gyromagnetic ratio	: 3.5295
Quad. moment	: - 0.47

Resonance Freq. (MHz) referred to ^1H TMS resonance at 100 MHz

13.194_2

Δν : 350 Hz

Spectral width (Hz)	: 20 000
Number of data points	: 1 K
Pulse	: 40 μs/ 90°
Number of scans	: 200 000
Repetition rate (s)	: 0.02
^1H decoupled	: -
Exp. filter (Hz)	: 200

^{139}La

Spin	: 7/2
Nat. abund (%)	: 99.91
Receptivity / ^{13}C	: $3.36\ 10^2$
Gyromagnetic ratio	: 3.7787
Quad. moment	: 0.21

Resonance Freq. (MHz) referred to ^1H TMS resonance at 100 MHz

14.125_6

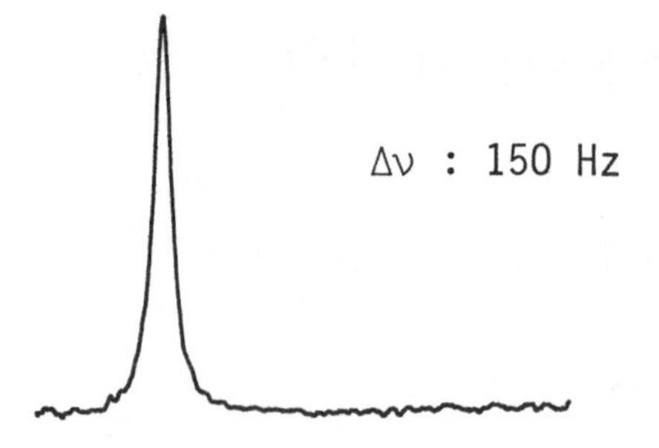

Δν : 150 Hz

Spectral width (Hz)	: 6 000
Number of data points	: 4 K
Pulse	: 40 μs/ 90°
Number of scans	: 1 000
Repetition rate (s)	: 0.34
^1H decoupled	: -
Exp. filter (Hz)	: 30

CHEMICAL SHIFTS : reference $LaCl_3$ - 10^{-2} M/D_2O

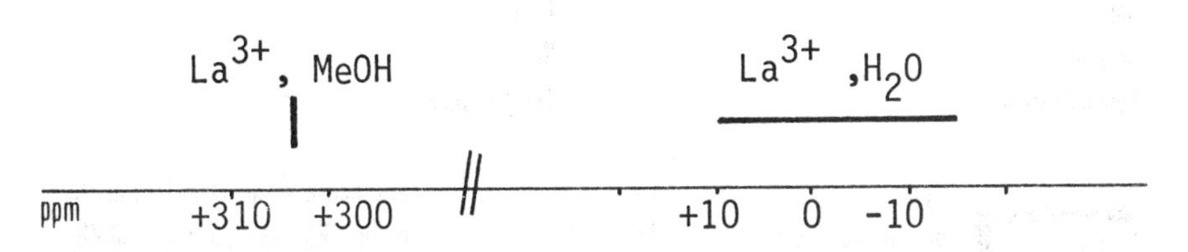

COUPLING CONSTANTS : typical values (Hz)

	^{1}H	^{19}F	^{31}P	^{13}C	Homo
^{1}J	- to -	- to -	- to -	- to -	- to -

RELAXATION :

T_1 typical values (s) : 10^{-3}

LITTERATURE :

B.12 p.200 ; B. 21 Vol. 10 A (1980)

O. Lutz, H. Oehler : J. Magn. Res., 37 261 (1980)

REMARKS : Linewidths are very solvent dependent

YTTERBIUM Z 70

Sample	: –	Lock	: –
Solvent	: –	Bo	: –
Concentration	: –	Temperature	: –

^{171}Yb

Spin	: 1/2
Nat abund (%)	: 14.27
Receptivity / ^{13}C	: 4.05
Gyromagnetic ratio	: (4.718)
Quad. moment	: –

Resonance Freq. (MHz) referred to ^{1}H TMS resonance at 100 MHz

(17.612)

no resonance found.
+II oxydation state
(diamagnetic) very
unstable

Spectral width (Hz)	: –
Number of data points	: –
Pulse	: – μs/
Number of scans	: –
Repetition rate (s)	: –
^{1}H decoupled	: –
Exp. filter (Hz)	: –

^{173}Yb

Spin	: 5/2
Nat. abund (%)	: 16.08
Receptivity / ^{13}C	: 1.14
Gyromagnetic ratio	: (1.310)
Quad. moment	: (0.4-3.9)

Resonance Freq. (MHz) referred to ^{1}H TMS resonance at 100 MHz

(4.852)

Spectral width (Hz)	: –
Number of data points	: –
Pulse	: – μs/
Number of scans	: –
Repetition rate (s)	: –
^{1}H decoupled	: –
Exp. filter (Hz)	: –

CHEMICAL SHIFTS : reference -

ppm

COUPLING CONSTANTS : typical values (Hz)

	^{1}H	^{19}F	^{31}P	^{13}C	Homo
^{1}J	- to -	- to -	- to -	- to -	- to -

RELAXATION :

T_1 typical values (s) : -

LITTERATURE :

L. Olschewski and al :
Z. Physik, 200, 224 (1967)

REMARKS : Only the +II oxydation state, although very instable, is diamagnetic

LUTETIUM Z 71

Sample	: –	Lock	: –
Solvent	: –	Bo	: –
Concentration	: –	Temperature	: –

^{175}Lu

Spin	: 7/2
Nat abund (%)	: 97.41
Receptivity / ^{13}C	: 156
Gyromagnetic ratio	: (3.05)
Quad. moment	: 5.68

Resonance Freq. (MHz) referred to ^1H TMS resonance at 100 MHz

(11.407)

no resonance found

Spectral width (Hz)	: –
Number of data points	: –
Pulse	: – μs/
Number of scans	: –
Repetition rate (s)	: –
^1H decoupled	: –
Exp. filter (Hz)	: –

^{176}Lu

Spin	: 7
Nat. abund (%)	: 2.59
Receptivity / ^{13}C	: 5.14
Gyromagnetic ratio	: (2.10)
Quad. moment	: 8.1

Resonance Freq. (MHz) referred to ^1H TMS resonance at 100 MHz

(7.872)

no resonance found

Spectral width (Hz)	: –
Number of data points	: –
Pulse	: – μs/
Number of scans	: –
Repetition rate (s)	: –
^1H decoupled	: –
Exp. filter (Hz)	: –

CHEMICAL SHIFTS : reference -

ppm

COUPLING CONSTANTS : typical values (Hz)

	^{1}H	^{19}F	^{31}P	^{13}C	Homo
^{1}J	- to -	- to -	- to -	- to -	- to -

RELAXATION :

T_1 typical values (s) : certainly below 10^{-6}

LITTERATURE :

I.J. Spalding and al : Proc. Phys. Soc. London 79, 787 (1962) (^{176}Lu)

G.J. Ritter and al : Phys. Rev. 126, 1493 (1962) (^{175}Lu)

REMARKS : useless for High Resolution NMR

HAFNIUM Z 72

Sample	: -		Lock	: -
Solvent	: -		Bo	: -
Concentration	: -		Temperature	: -

^{177}Hf

Spin	:	7/2
Nat abund (%)	:	18.50
Receptivity / ^{13}C	:	0.88
Gyromagnetic ratio	:	(+ 0.945)
Quad. moment	:	4.5

Resonance Freq. (MHz) referred to ^{1}H TMS resonance at 100 MHz

(4.008)

no resonance observed

Spectral width (Hz)	:	-	
Number of data points	:	-	
Pulse	:	-	µs/
Number of scans	:	-	
Repetition rate (s)	:	-	
^{1}H decoupled	:	-	
Exp. filter (Hz)	:	-	

^{179}Hf

Spin	:	9/2
Nat. abund (%)	:	13.75
Receptivity / ^{13}C	:	0.27
Gyromagnetic ratio	:	(- 0.609)
Quad. moment	:	5.1

Resonance Freq. (MHz) referred to ^{1}H TMS resonance at 100 MHz

(2.518)

no resonance observed

Spectral width (Hz)	:	-	
Number of data points	:	-	
Pulse	:	-	µs/
Number of scans	:	-	
Repetition rate (s)	:	-	
^{1}H decoupled	:	-	
Exp. filter (Hz)	:	-	

CHEMICAL SHIFTS : reference -

ppm

COUPLING CONSTANTS : typical values (Hz)

	^{1}H	^{19}F	^{31}P	^{13}C	Homo
^{1}J	- to -	- to -	- to -	- to -	- to -

RELAXATION :

T_1 typical values (s) : -

LITTERATURE :

S. Buettgenbach and al :

Z. Physik, 260, 157 (1973)
Phys. Letters, B. Vol. 43, 479 (1973)

REMARKS :

TANTALUM Z 73

Sample	:	$KTaCl_6$	Lock	:	unlocked
Solvent	:	CH_3CN	Bo	:	5.875
Concentration	:	saturated	Temperature	:	27°

^{181}Ta

Spin	:	7/2
Nat. abund (%)	:	99.988
Receptivity / ^{13}C	:	$2.04\ 10^2$
Gyromagnetic ratio	:	3.2073
Quad. moment	:	3

Resonance Freq. (MHz) referred to ^1H TMS resonance at 100MHz

11.989_6

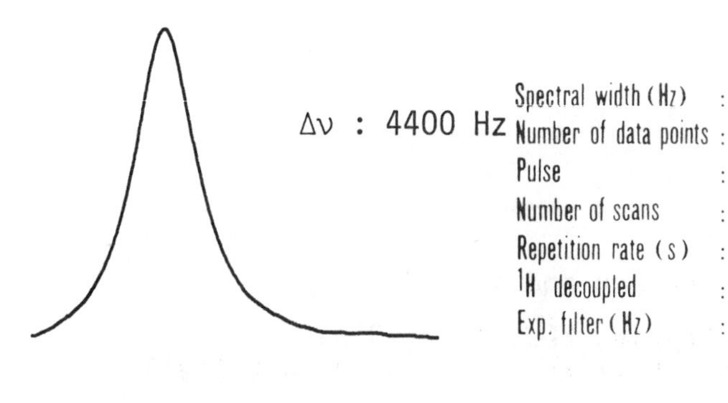

Spectral width (Hz)	:	50 000
Number of data points	:	2 K
Pulse	:	50 μs / 45°
Number of scans	:	2 000
Repetition rate (s)	:	0.02
^1H decoupled	:	-
Exp. filter (Hz)	:	500

CHEMICAL SHIFTS : reference $TaCl_6^-$

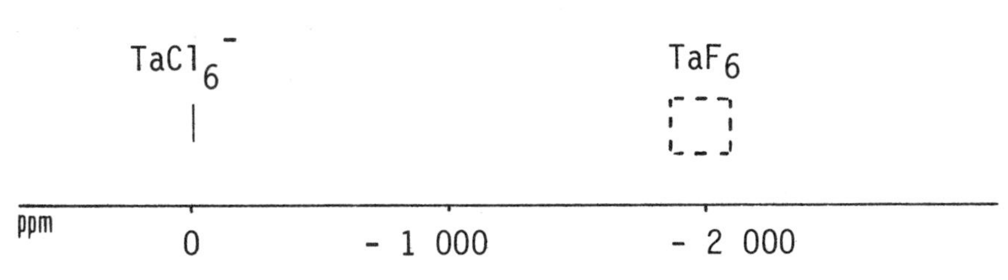

COUPLING CONSTANTS : typical values (Hz)

	^1H	^{19}F	^{31}P	^{13}C	Homo
^1J	- to -	- to -	- to -	- to -	- to -

RELAXATION :

T_1 typical values (s) : 3.10^{-5}

LITTERATURE :

B. 12 p. 212 ; B. 21 Vol. 10 A (1980)

L.C. Erich et al. : J. Chem. Phys., 59, 3911 (1973)

REMARKS : One of the highest quadrupole moment known. Very broad lines even for symmetrical environments.

TUNGSTEN Z 74

Sample	: Na_2WO_4	Lock	:	^2H
Solvent	: D_2O	Bo	:	5.875
Concentration	: 1 M	Temperature	:	27°

^{183}W

Spin	: 1/2
Nat. abund (%)	: 14.28
Receptivity / ^{13}C	: 5.89 10^{-2}
Gyromagnetic ratio	: 1.1145
Quad. moment	: -

Resonance Freq. (MHz) referred to ^1H TMS resonance at 100MHz

4.1663

*

Δν : 2 Hz

Spectral width (Hz)	: 2 000
Number of data points	: 4 K
Pulse	: 70 μs / 60°
Number of scans	: 3 200
Repetition rate (s)	: 1.02
^1H decoupled	: -
Exp. filter (Hz)	: 2

CHEMICAL SHIFTS : reference Na_2WO_4 M/D_2O

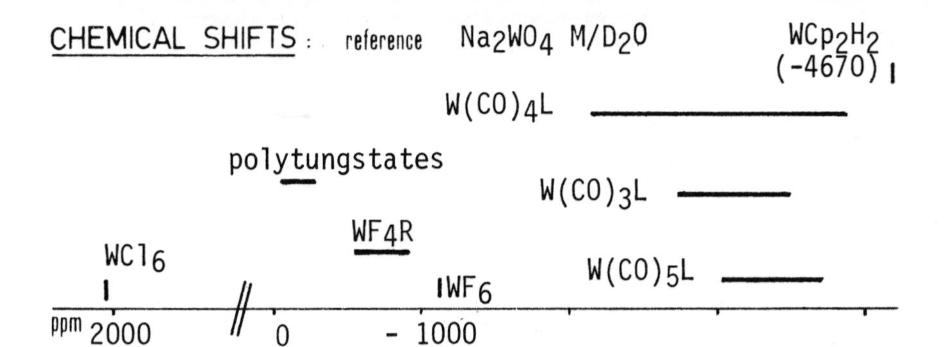

COUPLING CONSTANTS : typical values (Hz)

	^1H	^{19}F	^{31}P	^{13}C	Homo
$\lvert ^1J \rvert$	28 to 80	10 to 80	200 to 500	120 to 190	- to -

RELAXATION :

T_1 typical values (s) : 0.3 to 10

LITTERATURE :

B 12 p. 215 ; B. 21 Vol. 10 A (1980)

Acerete et al., J. Am. Chem. Soc., 101 267 (1979) ; J. Chem. Soc., Chem. Comm., 777 (1979)
O.A. Gansow et al., J. Organometal. Chem., 187, C27 (1980)

REMARKS : $^2J_{W-O-W}$ range from 2 to 26 Hz

RHENIUM Z 75

Sample	: $KReO_4$		Lock	: ^2H
Solvent	: D_2O		Bo	: 5.875
Concentration	: 0.1 M		Temperature	: 70°

^{185}Re

Spin	: 5/2
Nat abund (%)	: 37.07
Receptivity / ^{13}C	: $2.8\ 10^2$
Gyromagnetic ratio	: 6.0255
Quad. moment	: 2.8

Resonance Freq. (MHz) referred to ^1H TMS resonance at 100 MHz

22.524_6

$\Delta\nu$: 5 400 Hz

Spectral width (Hz)	: 50 000
Number of data points	: 1 K
Pulse	: 40 µs/ 90°
Number of scans	: 100 000
Repetition rate (s)	: 0.010
^1H decoupled	: -
Exp. filter (Hz)	: 300

^{187}Re

Spin	: 5/2
Nat. abund (%)	: 62.93
Receptivity / ^{13}C	: $4.90\ 10^2$
Gyromagnetic ratio	: 6.0862
Quad. moment	: 2.6

Resonance Freq. (MHz) referred to ^1H TMS resonance at 100 MHz

22.751_6

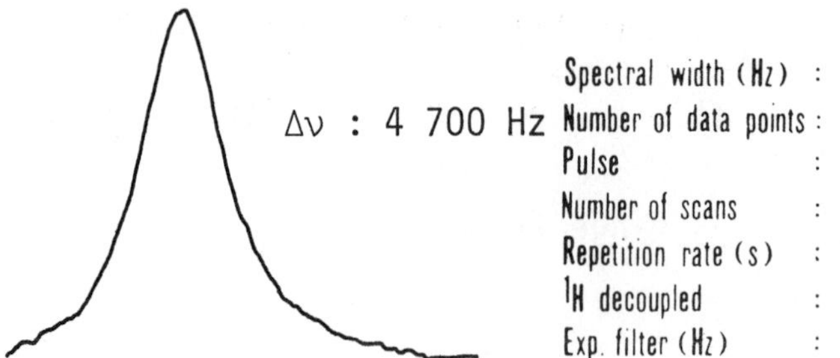

$\Delta\nu$: 4 700 Hz

Spectral width (Hz)	: 45 000
Number of data points	: 1 K
Pulse	: 35 µs/ 90°
Number of scans	: 100 000
Repetition rate (s)	: 0.011
^1H decoupled	: -
Exp. filter (Hz)	: 300

CHEMICAL SHIFTS : reference NaReO$_4$, aqueous

The reference sample provides the only observed resonance.

---

ppm

COUPLING CONSTANTS : typical values (Hz)

	^{1}H	^{19}F	^{31}P	^{13}C	Homo
^{1}J	- to -	- to -	- to -	- to -	- to -

RELAXATION :

T_1 typical values (s) : below 10^{-5}

LITTERATURE :

B. 12 p. 223 ; B. 21 Vol. 10 A (1980)

R.A. Dwek et al. : J. Phys. Chem., 74, 2232 (1970)

REMARKS : Difficult to observe (very broad lines), except for very symmetrical compounds. ^{187}Re is the most favored isotope.

OSMIUM Z 76

Sample	: OsO_4	Lock	: unlocked
Solvent	: CCl_4	Bo	: 5.875
Concentration	: 0.98 M	Temperature	: 27°

^{187}Os

Spin	: 1/2
Nat abund (%)	: 1.64
Receptivity / ^{13}C	: $1.14\ 10^{-3}$
Gyromagnetic ratio	: 0.6105
Quad. moment	: -

Resonance Freq. (MHz) referred to ^{1}H TMS resonance at 100 MHz

2.282_3

$\Delta\nu$: 10 Hz

Spectral width (Hz)	: 6 000
Number of data points	: 1 K
Pulse	: 100 μs/ 90°
Number of scans	: 150 000
Repetition rate (s)	: 0.17
^{1}H decoupled	: -
Exp. filter (Hz)	: 10

^{189}Os

Spin	: 3/2
Nat. abund (%)	: 16.1
Receptivity / ^{13}C	: 2.13
Gyromagnetic ratio	: 2.0773
Quad. moment	: 0.8

Resonance Freq. (MHz) referred to ^{1}H TMS resonance at 100 MHz

7.765_4

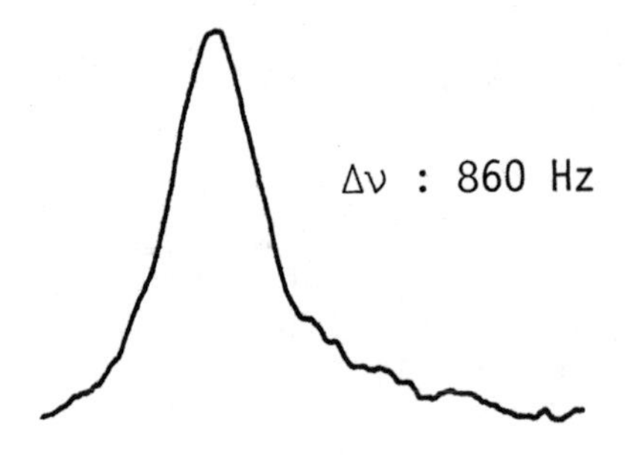

$\Delta\nu$: 860 Hz

Spectral width (Hz)	: 20 000
Number of data points	: 1 K
Pulse	: 60 μs/ 90°
Number of scans	: 30 000
Repetition rate (s)	: 0.02
^{1}H decoupled	: -
Exp. filter (Hz)	: 100

CHEMICAL SHIFTS : reference OsO_4

The only known resonance is from OsO_4

ppm

COUPLING CONSTANTS : typical values (Hz)

	^{1}H	^{19}F	^{31}P	^{13}C	Homo
^{1}J	- to -	- to -	- to -	- to -	- to -

RELAXATION :

T_1 typical values (s) :

$^{187}OsO_4$: between 1 and 20

$^{189}OsO_4$: 2 to 3 10^{-4}

LITTERATURE :

B.12 p. 224 ; B. 21 Vol. 10 A (1980)

Schwenk et al. : Z. Naturforsch., 29 a 1763 (1974)
Phys. Letters 26 A, 258 (1968)

REMARKS : Caution should be exercised when handling OsO_4

IRIDIUM Z 77

Sample	:	-	Lock	:	-
Solvent	:	-	Bo	:	-
Concentration	:	-	Temperature	:	-

^{191}Ir

Spin	:	3/2
Nat abund (%)	:	37.3
Receptivity / ^{13}C	:	0.023
Gyromagnetic ratio	:	(0.539)
Quad. moment	:	1.5

Resonance Freq. (MHz) referred to ^{1}H TMS resonance at 100 MHz

(1.718)

no resonance observed

Spectral width (Hz)	:	-	
Number of data points	:	-	
Pulse	:	-	μs/
Number of scans	:	-	
Repetition rate (s)	:	-	
^{1}H decoupled	:	-	
Exp. filter (Hz)	:	-	

^{193}Ir

Spin	:	3/2
Nat. abund (%)	:	62.7
Receptivity / ^{13}C	:	0.050
Gyromagnetic ratio	:	(0.391)
Quad. moment	:	1.4

Resonance Freq. (MHz) referred to ^{1}H TMS resonance at 100 MHz

(1.871)

no resonance observed

Spectral width (Hz)	:	-	
Number of data points	:	-	
Pulse	:	-	μs/
Number of scans	:	-	
Repetition rate (s)	:	-	
^{1}H decoupled	:	-	
Exp. filter (Hz)	:	-	

CHEMICAL SHIFTS : reference -

ppm

COUPLING CONSTANTS : typical values (Hz)

	^{1}H	^{19}F	^{31}P	^{13}C	Homo
^{1}J	- to -	- to -	- to -	- to -	- to -

RELAXATION :

T_1 typical values (s) : -

LITTERATURE :

N.V. Virgman and al : J. Chem. Phys. 59, 4418 (1973)

REMARKS : Very low resonance frequency for both isotopes. Useless for high resolution NMR

PLATINUM Z 78

Sample	: $Na_2(PtCl_4)$	Lock	: ^2H
Solvent	: D_2O	Bo	: 2.114
Concentration	: 1.2 M	Temperature	: 27°

^{195}Pt

Spin	: 1/2
Nat. abund (%)	: 33.8
Receptivity / ^{13}C	: 19.1
Gyromagnetic ratio	: 5.7412
Quad. moment	: -

Resonance Freq. (MHz) referred to ^1H TMS resonance at 100MHz

21.4617_4

$\Delta\nu$: 21 Hz

Spectral width (Hz)	: 1 000
Number of data points	: 4 K
Pulse	: 20 µs / 90°
Number of scans	: 10
Repetition rate (s)	: 2
^1H decoupled	: -
Exp. filter (Hz)	: 4

CHEMICAL SHIFTS : reference $Na_2 PtCl_6$

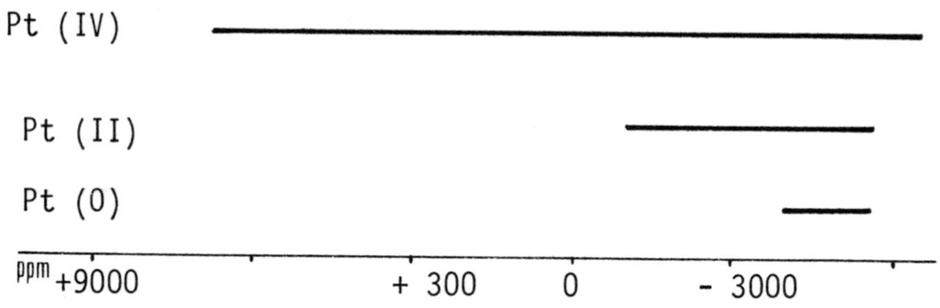

Pt (IV)

Pt (II)

Pt (0)

ppm +9000 + 300 0 - 3000

COUPLING CONSTANTS : typical values (Hz)

	^1H	^{19}F	^{31}P	^{13}C	Homo
$\lvert {}^1J \rvert$	+ 700 to + 1300	200 to 2000	+ 1080 to + 4510	- 5321 to + 906	+ 92 to + 9000

RELAXATION :

T_1 typical values (s) : 1.3 to 0.3

LITTERATURE :

B. 12 p. 249

S.J.S. Kerrison and P.J. Sadler, J. Magn. Reson., 31, 321 (1978)

R.G. Kidd "Nuclear Shielding of the Transitions Metals" (B 21), Vol. 10 A, (1980)

REMARKS : Large temperature effect on δ (0.5 to 1.1 ppm/°K). δ and J are structure dependent.
Minimum concentration : 10^{-2}M.

GOLD Z 79

Sample	:	-	Lock	:	-
Solvent	:	-	Bo	:	-
Concentration	:	-	Temperature	:	-

^{197}Au

Spin	:	3/2
Nat. abund (%)	:	100
Receptivity / ^{13}C	:	0.06
Gyromagnetic ratio	:	(0.357)
Quad. moment	:	0.58

Resonance Freq. (MHz) referred to ^{1}H TMS resonance at 100MHz

(1.729)

no resonance observed

Spectral width (Hz)	:	-	
Number of data points	:	-	
Pulse	:	-	μs /
Number of scans	:	-	
Repetition rate (s)	:	-	
^{1}H decoupled	:	-	
Exp. filter (Hz)	:	-	

CHEMICAL SHIFTS : reference -

ppm

COUPLING CONSTANTS : typical values (Hz)

	^1H	^{19}F	^{31}P	^{13}C	Homo
^1J	- to -	- to -	- to -	- to -	- to -

RELAXATION :

T_1 typical values (s) : -

LITTERATURE :

A. Narath : Phys. Review, 163, 232 (1967)
175, 696 (1968) (metal)

H. Dahnen and al. : Z. Physik 200, 456 (1967)

REMARKS : Very low frequency. This nucleus should be very difficult to detect.

MERCURY Z 80

Sample	: $HgMe_2$	Lock	: ^2H
Solvent	: C_6D_6	Bo	: 5.875
Concentration	: 90/10 (v/v)	Temperature	: 27

^{199}Hg

Spin	: 1/2
Nat abund (%)	: 16.84
Receptivity / ^{13}C	: 5.42
Gyromagnetic ratio	: 4.7912
Quad. moment	: -

Resonance Freq. (MHz) referred to ^1H TMS resonance at 100 MHz

17.910_7

$\Delta\nu$: 1 Hz

Spectral width (Hz)	: 4 000
Number of data points	: 8 K
Pulse	: 25 µs/ 40°
Number of scans	: 1
Repetition rate (s)	: -
^1H decoupled	: yes
Exp. filter (Hz)	: 1

^{201}Hg

Spin	: 3/2
Nat. abund (%)	: 13.22
Receptivity / ^{13}C	: 1.08
Gyromagnetic ratio	: -1.7686
Quad. moment	: 0.5

Resonance Freq. (MHz) referred to ^1H TMS resonance at 100 MHz

(6.611_5)

Line too broad to be observed.
$^{201}\gamma/^{199}\gamma$ = - 0.369139
(Compt.Rend.Acad. : 285, B p. 45)

Spectral width (Hz)	: -
Number of data points	: -
Pulse	: - µs/
Number of scans	: -
Repetition rate (s)	: -
^1H decoupled	: -
Exp. filter (Hz)	: -

CHEMICAL SHIFTS: reference $Hg(Me)_2$ neat

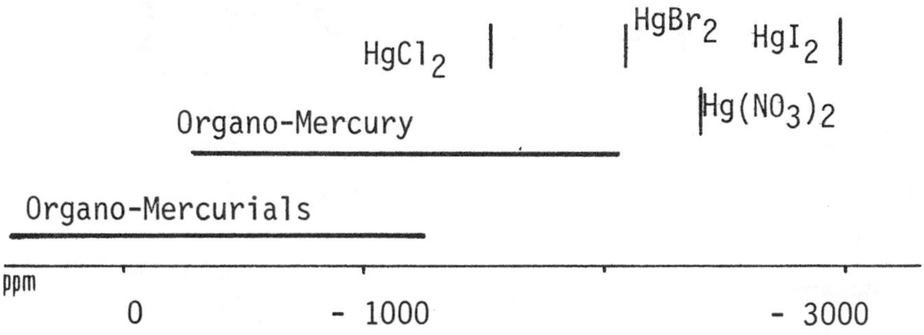

$HgCl_2$ | $HgBr_2$ | HgI_2 | $Hg(NO_3)_2$

Organo-Mercury

Organo-Mercurials

ppm

0 — -1000 — -3000

COUPLING CONSTANTS: typical values (Hz)

	^1H	^{19}F	^{31}P	^{13}C	Homo
$\lvert ^1J \rvert$ (^{199}Hg)	- to -	to	3 000 to 7 500	500 to 1 800	- to -

RELAXATION :

T_1 typical values (s) : 1 to 10^{-2}

LITTERATURE :

B. 12 p. 266

B. 21 Vol. 10 A (1980)

REMARKS: $^2J_{Hg-H}$ range from 100 to 300 Hz

$^2J_{Hg-F}$ range from 1 000 to 2 000 Hz

THALLIUM Z 81

Sample	: $Tl(NO_3)_3$		Lock	: ^2H
Solvent	: D_2O/HNO_3		Bo	: 5.875
Concentration	: 0.5 M		Temperature	: 27°

^{203}Tl

Spin	: 1/2
Nat abund (%)	: 29.50
Receptivity / ^{13}C	: 2.89 10^2
Gyromagnetic ratio	: 15.3078
Quad. moment	: -

Resonance Freq. (MHz) referred to ^1H TMS resonance at 100 MHz

57.223_6

$\Delta\nu$: 35 Hz

Spectral width (Hz)	: 6 000
Number of data points	: 8 K
Pulse	: 60 μs/ 90°
Number of scans	: 10
Repetition rate (s)	: 1.6
^1H decoupled	: -
Exp. filter (Hz)	: 5

^{205}Tl

Spin	: 1/2
Nat. abund (%)	: 70.50
Receptivity / ^{13}C	: 7.69 10^2
Gyromagnetic ratio	: 15.4584
Quad. moment	: -

Resonance Freq. (MHz) referred to ^1H TMS resonance at 100 MHz

57.786_4

$\Delta\nu$: 35 Hz

Spectral width (Hz)	: 6 000
Number of data points	: 8 K
Pulse	: 60 μs/ 90°
Number of scans	: 10
Repetition rate (s)	: 1.6
^1H decoupled	: -
Exp. filter (Hz)	: 5

CHEMICAL SHIFTS : reference $TlNO_3$, H_2O, (infinite dilution)

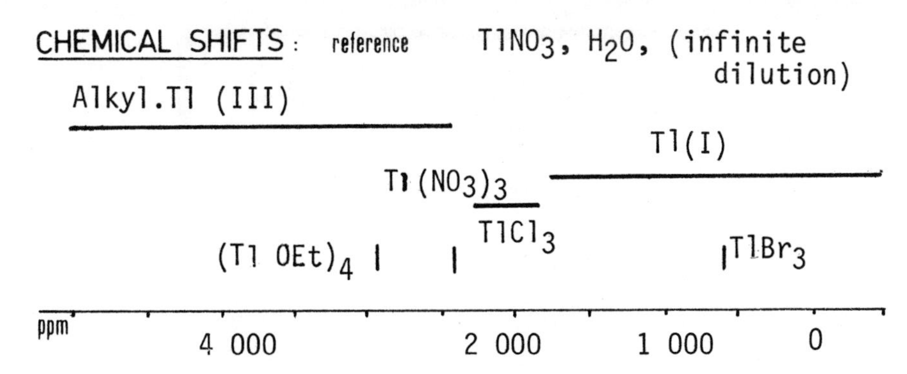

COUPLING CONSTANTS : typical values (Hz)

	^1H	^{19}F	^{31}P	^{13}C	Homo
$\lvert ^1J \rvert$ (^{205}Tl)	6 000 to (estimated)	to	to	1 900 to 10 500	to

RELAXATION :

T_1 typical values (s) : ^{203}Tl, ^{205}Tl : 10^{-1} to 2

LITTERATURE :

B. 12 p. 288

REMARKS : Chemical shifts are extremely solvent, concentration and temperature dependent

LEAD Z 82

Sample	: $Pb(NO_3)_2$		Lock	: ^2H
Solvent	: D_2O		Bo	: 2.114
Concentration	: saturated		Temperature	: 27°

^{207}Pb

Spin	: 1/2
Nat. abund (%)	: 22.6
Receptivity / ^{13}C	: 11.8
Gyromagnetic ratio	: 5.5797
Quad. moment	: -

Resonance Freq. (MHz) referred to ^1H TMS resonance at 100MHz

20.858_1

$\Delta\nu$: 3.5 Hz

Spectral width (Hz)	: 2 000
Number of data points	: 4 K
Pulse	: 12 μs / 70°
Number of scans	: 550
Repetition rate (s)	: 1 024
^1H decoupled	: -
Exp. filter (Hz)	: 2

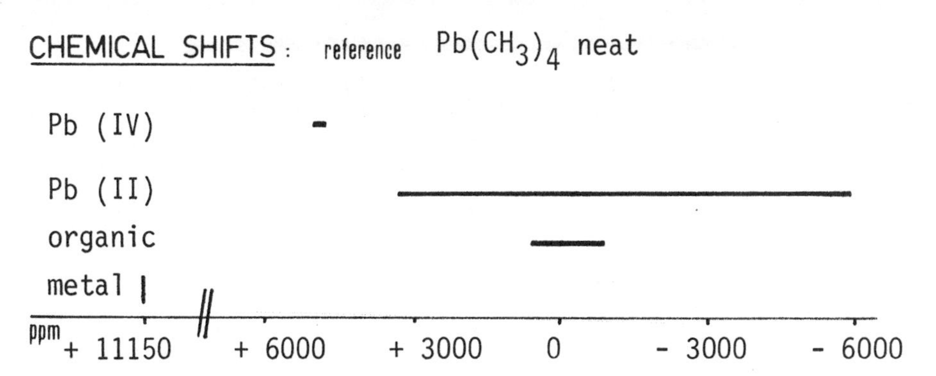

COUPLING CONSTANTS : typical values (Hz)

	^{1}H	^{19}F	^{31}P	^{13}C	Homo
$\lvert ^{1}J \rvert$	2454	- to -	- to -	28 to 16625	- to -

RELAXATION :

T_1 typical values (s) : 2 to 0.1

LITTERATURE :

B. 12 p. 366

REMARKS : The observation of this nucleus is easy.
Small solvent effects.

BISMUTH Z 83

Sample	: $Bi(NO_3)_3$	Lock	: C_6F_6
Solvent	: HNO_3 conc.	Bo	: 2.114
Concentration	: saturated	Temperature	: 27°

^{209}Bi

Spin	: 9/2
Nat. abund (%)	: 100
Receptivity / ^{13}C	: 777
Gyromagnetic ratio	: 4.2986
Quad. moment	: -0.4

Resonance Freq. (MHz) referred to ^1H TMS resonance at 100MHz

16.069_2

$\Delta\nu$: 3 200 Hz

Spectral width (Hz)	: 20 000
Number of data points	: 1 K
Pulse	: 15 μs / 80°
Number of scans	: 10 000
Repetition rate (s)	: 0.025
^1H decoupled	: -
Exp. filter (Hz)	: 50

CHEMICAL SHIFTS : reference $KBiF_6$

Only $Bi(NO_3)_3$ and $KBiF_6$ has been observed

ppm

COUPLING CONSTANTS : typical values (Hz)

	^1H	^{19}F	^{31}P	^{13}C	Homo
$\|^1J\|$	to	2 700 (BiF_6^-)	to	to	to

RELAXATION :

T_1 typical values (s) : below 10^{-4}

LITTERATURE :

B. 12 p. 381

REMARKS : Large quadrupole moment prevents the observation of Bi resonances in non symmetrical compounds.

REFERENCES

BOOKS

Only recent books or those of interest to the NMR spectroscopist are cited.

B 1. A. Abragam, (a) The Principles of Nuclear Magnetism, Oxford University Press, London, 1961; (b) Les Principes du Magnétisme Nucléaire, Presses Universitaires de France, 1961.

B 2. R. J. Abraham, The Analysis of High Resolution NMR Spectra, Academic, New York, 1967.

B 3. I. V. Aleksandrov, The Theory of N.M.R., Academic, New York, 1966.

B 4. T. Axenrod and G. A. Webb, NMR of Nuclei other than Protons, Wiley, New York, 1974 (Meeting at Pisa, 1972).

B 5. L. J. Berliner and J. Reuben, Biological Magnetic Resonance Vol. 1, 1978; Vol. 2, 1979, Plenum, London.

B 6. A. Carrington and A. D. McLachlan, Introduction to Magnetic Resonance, Chapman and Hall, London, 1979.

B 7. P. L. Corio, Structure of High Resolution NMR Spectra Academic, New York, 1966.

B 8. P. Diehl, E. Fluck, and R. Kostfeld, Eds., NMR Basic Principles and Progress, Springer, Berlin. Vol. 1, 1969, to Vol. 17, 1981 (extending series).

B 9. J. W. Emsley, J. Feeney, and L. H. Sutcliffe, High Resolution NMR Spectroscopy, Vols. I and II, Pergamon, London, 1965.

B10. J. W. Emsley, J. Feeney, and L. H. Sutcliffe, Eds, Progress in NMR Spectroscopy, Pergamon, London Vol. 1, 1966 to Vol. 9, 1973 (extending series).

B11. T. C. Farrar and E. D. Becker, Pulse and Fourier Transform NMR, Academic, New York, 1971.

B12. R. K. Harris and B. E. Mann, NMR and the Periodic Table, Academic, New York, 1978.

B13. L. M. Jackman and F. A. Cotton, Eds, Dynamic NMR Spectroscopy, Academic, New York, 1975.

B14. F. Kasler, Quantitative Analysis by NMR Spectroscopy, Academic, New York, 1973.

B15. R. Lenk, Brownian Motion and Spin Relaxation, Elsevier, Amsterdam, 1977.

B16. G. C. Levy, Ed., Topics in 13C NMR Spectroscopy, Vol.1, 1974, to Vol. 3, 1979, Wiley, New York.

B17. G. C. Levy and R. L. Lichter, Nitrogen 15 NMR Spectroscopy, Wiley, New York, 1979.

B18. D. E. Leyden and R. H. Cox, Analytical Applications of NMR, Wiley, New York, 1977.

B19. W. McFarlane and R. F. M. White, Techniques of High Resolution NMR Spectroscopy, Butterworth, London, 1972.

B20. M. L. Martin, J. J. Delpuech, and G. J. Martin, Practical NMR Spectroscopy, Heyden, London, 1979.

B21. E. F. Mooney and G. A. Webb, Annual Reports of NMR Spectroscopy, Vol. 1, 1968, to Vol. 11, 1980, Academic, London (extending series).

B22. K. Müllen and P. S. Pregosin, Fourier Transform NMR Techniques - A Practical Approach, Academic, New York, 1976.

B23. J. H. Noggle and R. E. Schirmer, The Nuclear Overhauser Effect - Chemical Applications, Academic, New York, 1971.

B24. H. Nöth and B. Wrackmeyer, NMR Spectroscopy of Boron Compounds, NMR Basic Principle and Progress, Vol. 14, Springer, Berlin, 1978.

B25. C. P. Poole, J. Horacio, and A. Farach, The Theory of Magnetic Resonance, Wiley, New York, 1972.

B26. J. A. Pople, W. G. Schneider, and H. S. Bernstein, High Resolution NMR, McGraw-Hill, New York, 1959.

B27. P. S. Pregosin and R. W. Kunz, 31P and 13C NMR of Transition Metal Phosphine Complexes, Springer, Berlin, 1979.

B28. D. Shaw, Fourier Transform NMR Spectroscopy, Elsevier, Amsterdam, 1976.

B29. R. E. Sievers, Nuclear Magnetic Shift Reagents, Academic, New York, 1973.

B30. C. P. Slichter, Principle of Magnetic Resonance, 2nd ed., Springer, New York, 1978.

B31. J. S. Waugh, Ed., Advances in Magnetic Resonance, Vol. 1, 1965, to Vol. 9, 1977, Academic, New York (extending series).

B32. F. W. Wehrli and T. Wirthlin, Interpretation of 13C NMR Spectra, Heyden, London, 1976.

B33. M. Witanowski and G. A. Webb, Nitrogen NMR, Plenum London, 1973.

B34. Nuclear Magnetic Resonance, The Chemical Society, London, ed., Vol. 1, 1972, Periodical Reports, Alden et Roubray, Oxford.

PERIODICALS

1. Pure Appl. Chem., 45, 217 (1976).

2. Org. Magn. Reson., 11, 267 (1978).

3. M. Suzuki and R. Kubo, Mol. Phys., 7, 201, (1964).

4. G. J. Schrobilgen, J. H. Holloway, P. Granger, and C. Brevard, Inorg. Chem., 17, 980 (1978).

5. C. Brown, B. T. Heaton, P. Chini, A. Fuornazalli, and G. Longoni, J. Chem. Soc. Chem. Commun., 309 (1977).

6. M. Rubinstein, A. Baram, and Z. Luz, Mol. Phys., 20 67 (1971).

7. J. Y. Lallemand, J. Soulié, and J. C. Chottard, J. Chem. Soc. Chem. Comm. 436 (1980).

8. N. F. Ramsey, Phys. Rev., 78, 699 (1950); C. Deverell, Mol. Phys., 18, 319 (1970).

9. P. Granger and S. Chapelle, J. Magn. Reson., 39, 329 (1980).

10. E. A. C. Lucken, Nuclear Quadrupole Coupling Constants, Academic, New York, 1969.

11. R. R. Ernst and W. A. Anderson, Rev. Sci. Instr., 37, 93 (1966).

12. J. Homer, A. R. Dudley, and W. R. McWhinnie, J. Chem. Soc. Chem. Commun., 893 (1973).

13. P. Stilbs and M. E. Moseley, Chem. Scr. 13, 26 (1978-1979).

14. J. K. Becconsall, G. D. Daves, and W. R. Anderson, J. Am. Chem. Soc., 92, 430 (1970).

15. D. H. Live and S. I. Chan, Anal. Chem., 42, 791 (1970).

16. W. McFarlane, Proc. R. Soc. London, Ser. A, 306, 185 (1968).

17. R. K. Harris and B. J. Kimber, J. Magn. Reson., 17, 174 (1975).

18. S. Brownstein and J. Bornais, J. Magn. Reson., 38, 131 (1980).

19. C. J. Turner, Varian Application Topic-9, October 1979.

20. H. Bernstein and K. Frei, J. Chem. Phys., 37, 1891 (1962).

21. N. C. Li, R. L. Scruggs, and E. D. Becker, J. Am. Chem. Soc., 84, 4650 (1962).

22. D. C. Douglass and A. Fratiello, J. Chem. Phys., 39, 3163 (1963).

23. D. F. Evans, Proc. Chem. Soc. London, 115 (1958).

24. D. F. Evans, J. Chem. Soc., 2003 (1959).

25. R. A. Bailey, J. Chem. Educ., 49, 297 (1972).

26. J. Q. Adams, Rev. Sci. Instrum. 37, 1099 (1966).

27. D. A. Gansow, K. M. Triplette, T. T. Peterson, R. E. Botto, and J. D. Roberts, Org. Magn. Reson., 13, 77 (1980).

28. S. Schäublin, A. Hohener, and R. R. Ernst, J. Magn. Reson., 13, 196 (1977).

29. A. Demarco and K. Wüthrich, J. Magn. Reson., 24, 201 (1976).

30. B. Clin, J. de Bony, P. Lalanne, J. Biais, and B. Lemanceau, J. Magn. Reson., 33, 457, (1979).

31. M. Guéron, J. Magn. Reson., 30, 515 (1978).

32. A. G. Ferrige and J. C. Lindon, J. Magn. Reson., 31, 337 (1978).

33. J. C. Lindon and A. G. Ferrige, J. Magn. Reson., 36, 277 (1979).

34. I. D. Campbell, C. M. Dobson, R. J. Williams, and A. V. Xavier, J. Magn. Reson., 11, 172 (1973).

35. J. W. Akitt, J. Magn. Reson., 32, 311 (1979).

36. (a) M. Taquin, Compt. Rend. Ac. Sci., 280B, 485 (1975).

(b) M. Taquin, Compt. Rend. Ac. Sci., 283B, 257 (1976).

37. K. G. R. Pachler and P. L. Wessels, J. Magn. Reson., 12, 337 (1973).

38. N. J. Koole and M. J. A. De Bie, J. Magn. Reson., 23, 9 (1976).

39. K. G. R. Pachler and P. L. Wessels, J. Magn. Reson., 28, 53 (1977).

40. F. W. Dahlquist, K. J. Longmuir, and R. B. Du Vernet, J. Magn. Reson., 17, 406 (1975).

41. C. L. Mayne, D. W. Alderman, and D. M. Grant, J. Chem. Phys., 63, 2514 (1975).

42. S. A. Linde, H. S. Jakobsen, and B. J. Kimber, J. Am. Chem. Soc. 97, 3219 (1975).

43. S. A. Linde and H. S. Jakobsen, J. Am. Chem. Soc., 98, 1041 (1976).

44. J. P. Jacobsen, J. Magn. Reson., 37, 365 (1980).

45. J. P. Jesson, P. Meakin, and G. Kneissel, J. Am. Chem. Soc. 95, 618 (1973).

46. W. A. Anderson and R. Freeman, J. Chem. Phys., 37, 85 (1962).

47. J. Feeney and P. Partington, J. Chem. Soc. Chem. Commun, 611 (1973).

48. A. A. Chalmers, K. G. R. Pachler, and P. L. Wessels, J. Magn. Reson., 15, 419 (1974).

49. S. Sorensen, R. S. Hansen, and H. J. Jakobsen, J. Chem. Phys., 52, 1529 (1970).

50. K. Kushida, K. Aoki, and S. Satoh, J. Am. Chem. Soc, 97, 44 (1975).

51. K. Bock, R. Burton, and L. D. Hall, Can. J. Chem., 54, 3526 (1976).

52. K. G. R. Pachler and P. L. Wessels, J. Chem. Soc. Chem. Commun., 1038 (1974).

53. E. L. Hahn, Phys. Rev., 80, 580 (1950).

54. P. Mansfield, Phys. Rev. 137, A 961 (1965).

55. J. H. Davis, K. R. Jeffrey, M. Bloom, M. I. Valic, and T. P. Higgs, Chem. Phys. Lett., 42, 390 (1976).

56. B. L. Tomlinson and H. D. W. Hill, J. Chem. Phys., 59, 1775 (1973).

57. I. D. Campbell, C. M. Dobson, and G. W. Jeminet, FEBS Letters, 49, 115 (1974).

58. J. Feeney and G. C. K. Roberts, J. Chem. Soc. Chem. Commun., 205 (1971).

59. S. L. Pattand and D. B. Sykes, J. Chem. Phys., 56, 3182 (1972).

60. K. Roth, Org. Magn. Reson., 12, 271 (1979).

61. T. R. Krugh and W. C. Schaefer, J. Magn. Reson., 19, 99 (1975).

62. D. W. Lowman and G. E. Maciel, Anal. Chem., 51, 85 (1979).

63. R. K. Harris, R. H. Newman, and A. Okruszek, Org. Magn. Res., 9, 58 (1977).

64. G. Bodenhausen, R. Freeman, and G. A. Morris, J. Magn. Reson., 23, 171 (1976).

65. A. Bax, A. F. Mehlkopf, and J. Smidt, J. Magn. Reson., 35, 373 (1979).

66. L. D. Hall and S. Sukumar, J. Magn. Reson., 38, 559 (1980).

67. J. A. B. Lohman, J. Magn. Reson., 38, 163 (1980).

68. D. Canet, J. P. Marchal, and J. P. Sartaux, Compt. Rend. Ac. Sci., 279B, 71 (1974).

69. J. P. Marchal and D. Canet, J. Am. Chem. Soc., 97, 6581 (1975).

70. A. I. Scott, M. Kajiwara, T. Takahashi, I. M. Armitage, P. Demon, and D. Petrocine, J. Chem. Soc. Chem. Commun., 544 (1976).

71. J. Hochmann and H. P. Kellerhals, J. Magn. Reson., 38, 23 (1980).

72. G. A. Morris and R. Freeman, J. Magn. Reson., 29, 433 (1978).

73. S. Forsen and R. A. Hoffman, J. Chem. Phys., 39, 2892 (1963); 40, 1189 (1964).

74. F. W. Dahlquist, K. J. Longmuir, and R. B. du Vernet, J. Magn. Reson., 17, 406 (1975).

75. E. D. Becker, J. A. Ferretti, and T. C. Farrar, J. Am. Chem. Soc., 91, 7784 (1969).

76. J. S. Waugh, J. Mol. Spectrosc., 35, 298 (1970).

77. R. R. Shoup, E. D. Becker, and T. C. Farrar, J. Magn. Reson., 8, 298 (1972).

78. (a) A. Allerhand and D. C. Cochran, J. Am. Chem. Soc., 92, 4482 (1970); (b) A. Allerhand, Rev. Sci. Instrum., 41, 269 (1970).

79. H. Y. Carr and E. M. Purcell, Phys. Rev., 94, 630 (1954).

80. G. G. McDonald and J. S. Leigh, J. Magn. Reson., 9, 358 (1973).

81. (a) M. H. Levitt and R. Freeman, J. Mag. Res., 33, 473 (1979); (b) R. Freeman, S. P. Kempsell, and M. H. Levitt, J. Magn. Reson., 38, 453 (1980).

82. R. O. Duthaler and J. D. Roberts, J. Am. Chem. Soc., 100, 4969 (1978).

83. H. J. Jakobsen and W. S. Brey, J. Am. Chem. Soc., 101, 774 (1979).

84. G. A. Morris and R. Freeman, J. Am. Chem. Soc., 101, 760 (1979).

85. R. D. Bertrand, W. B. Moniz, A. N. Garroway, and G. C. Chingas: (a) J. Am. Chem. Soc., 100, 5227 (1978); (b) J. Magn. Reson., 32, 465 (1978).

86. S. R. Hartmann and E. L. Hahn, Phys. Rev., 128, 2042 (1962).

87. B. Thiault and M. Mersseman, Org. Magn. Reson., 7, 575 (1975).

88. B. Thiault and M. Mersseman, Org. Magn. Reson., 8, 28 (1976).

89. S. Barcza and N. Engstron, J. Am. Chem. Soc., 94, 1762 (1972).

90. O. A. Gansow, A. R. Burke, and W. D. Vernon, J. Am. Chem. Soc., 94, 2550 (1972).

91. S. Gillet and J. J. Delpuech, J. Magn. Reson., 38, 433 (1980).

92. J. J. Led and S. B. Petersen, J. Magn. Reson., 32, 1 (1978).

93. K. Bock, B. Meyer, and M. Vignon, J. Magn. Reson., 38, 545 (1980).

94. T. G. Kollie, R. L. Anderson, J. L. Horton, and M. J. Roberts, Rev. Sci. Instrum., 48, 501 (1977).

95. R. A. Newmark and R. E. Graves, J. Phys. Chem., 72, 4299 (1968).

96. J. Bornais and S. Bronstein, J. Magn. Reson., 29, 207 (1978).

97. P. M. Henrichs and P. E. Peterson, J. Am. Chem. Soc., 95, 7449 (1973).

98. H. J. Schneider, W. Freitag, and M. Schommer, J. Magn. Reson., 18, 393 (1975).

99. D. W. Vidrine and P. E. Peterson, Anal. Chem., 48, 130 (1976).

100. G. C. Levy, J. T. Bailey, and D. A. Wright, J. Magn. Reson., 37, 353 (1980).

101. O. Yamamoto and M. Yanagisawa, Anal. Chem., 42, 1463 (1970).

102. M. L. Kaplan, F. A. Bovey, and H. N. Cheng, Anal. Chem., 47, 1703 (1975).

103. C. Piccini-Leopardi, O. Fabre, and J. Reisse, Org. Magn. Reson. 8, 233 (1976).

104. D. S. Stephenson and G. Binsch, J. Magn. Reson., 37, 395 and 409 (1980).

105. (a) J. Kaufmann and A. Schwenk, Phys. Lett. A24, 115 (1967); (b) A. Schwenk, Z. Phys. 213, 482 (1968).

106. A. Schwenk, J. Magn. Reson., 5, 376 (1971).

107. R. Freeman and H. D. W. Hill, J. Magn. Reson., 4, 366 (1971).

108. G. E. Chapman, B. D. Abercrombie, P. D. Gary, and E. M. Bradbury, J. Magn. Reson., 31, 459 (1978).

109. S. J. Opella, D. J. Nelson, and O. Jardetzky, J. Chem. Phys., 64, 2533 (1976).

110. K. F. Kuhlmann and D. M. Grant, J. Chem. Phys., 55, 2998 (1971).

111. D. Canet, J. Magn. Reson., 23, 361 (1976).

112. R. K. Harris and R. H. Newman, J. Magn. Reson., 24, 449 (1976).

113. P. Granger, S. Chapelle, and C. Brevard, J. Magn. Reson., 42, 203 (1981).

114. P. Granger, S. Chapelle, and J. M. Poirier, Org. Magn. Reson., 14, 69 (1980).

115. S. L. Patt, Application Topic. Varian N°. 79 (1977).

116. H. S. Gutowsky and D. F. S. Natush, J. Chem. Phys., 57, 1203 (1972).

117. S. L. Gordon and K. Wüthrich, J. Am. Chem. Soc. (in press).

118. G. Wagner and K. Wüthrich, J. Magn. Reson., 33, 675 (1979).

119. R. Freeman, H. D. W. Hill and R. Kaptein, J. Magn. Reson., 7, 327 (1972).

120. J. Kowalewski, A. Ericsson, and R. Vestin, J. Magn. Reson., 31, 165 (1978).

121. J. C. Duplan, A. Briguet, G. Tetu, and J. Delmau, J. Magn. Reson., 31, 509 (1978).

122. P. E. Fagerness, D. M. Grant, and R. B. Parry, J. Magn. Reson., 26, 267 (1977).

123. B. J. Kimber and R. K. Harris, J. Magn. Reson., 16, 354 (1974).

124. R. L. Vold, J. S. Waugh, M. P. Klein, and D. E. Phelps, J. Chem. Phys., 48, 3831 (1968).

125. R. Freeman and H. D. W. Hill, J. Chem. Phys., 51, 3140 (1969).

126. D. E. Demco, P. van Hecke, and J. S. Waugh, J. Magn. Reson., 16, 467 (1974).

127. D. Canet, G. C. Levy, and I. R. Peat, J. Magn. Reson., 18, 199 (1975).

128. J. D. Cutnell, H. E. Bleich, and J. A. Glasel, J. Magn. Reson., 21, 43 (1976).

129. J. Kowalewski, G. C. Levy, L. F. Johnson, and L. Palmer, J. Magn. Reson., 26, 533 (1977).

130. R. K. Gupta, J. A. Ferretti, E. D. Becker, and G. H. Weiss : J. Magn. Reson., 38, 447 (1980).

131. D. L. Rabenstein, T. Nakashima, and G. Bigam, J. Magn. Reson., 34, 669 (1979).

132. J. L. Markley, W. J. Horsley, and P. M. Klein, J. Chem. Phys., 55, 3604 (1971).

133. G. G. McDonald and J. S. Leigh, J. Magn. Reson., 9, 358 (1973).

134. W. Dietrich, G. Bergmann, and R. Gerhards, Z. Anal. Chem., 279, 177 (1976).

135. R. Freeman and H. D. W. Hill, J. Chem. Phys., 54, 3367 (1971).

136. K. A. Christensen, D. M. Grant, E. M. Schulman and C. Walling, J. Phys. Chem., 78, 1971 (1974).

137. Y. N. Luzikov, N. M. Sergeyev, and M. G. Lerkovitch, J. Magn. Reson., 21, 359 (1974).

138. R. K. Gupta, J. Magn. Reson., 25, 231 (1977).

139. R. K. Gupta, G. H. Weiss, J. A. Ferretti, and E. D. Becker, J. Magn. Reson., 35, 301 (1979).

140. R. Kaptein, K. Dijkstra, and C. E. Tarr, J. Magn. Reson., 24, 295 (1976).

141. H. Hanssum, W. Maurer, and H. Rüterjans, J. Magn. Reson., 31, 231 (1978).

142. M. Sass and D. Ziessow, J. Magn. Reson., 25, 263 (1977).

143. R. Gerhards and W. Dietrich, J. Magn. Reson., 23, 21 (1976).

144. F. W. Wehrli, Application Note Varian N°. NMR 77-2 (1977).

145. (a) G. H. Weiss, R. K. Gupta, J. B. Ferretti, and E. D. Becker, J. Magn. Reson., 37, 369 (1980)., (b) E. D. Becker, J. A. Ferretti, R. K. Gupta, and G. H. Weiss, J. Magn. Reson., 37, 381 (1980).

146. A. Kumar, C. S. Johnson, J. Magn. Reson., 7, 55 (1972).

147. S. Meiboom and D. Gill, Rev. Sci. Instrum., 29, 688 (1958).

148. D. G. Hughes, J. Magn. Reson., 26, 481 (1977).

149. R. L. Vold, R. R. Vold, and H. E. Simon, J. Magn. Reson., 11, 283 (1973).

150. T. E. Bull, Rev. Sci. Instrum., 45, 232 (1974).

151. D. G. Hughes and G. Lindblom, J. Magn. Reson., 26, 469 (1977).

152. E. L. Hahn and D. E. Maxwell, Phys. Rev., 88, 1070 (1952).

153. R. Freeman, H. D. W. Hill, and J. Daddok, J. Chem. Phys., 58, 3107 (1973).

154. R. Freeman and H. D. W. Hill, J. Chem. Phys., 54, 301 (1971).

155. J. Kronenbitter and A. Schwenk, J. Magn. Reson., 25, 147 (1977).

156. M. Eisenstadt, J. Magn. Reson., 38, 507 (1980).

157. D. Rogers, M. T. Rogers, G. D. Vickers, Rev. Sci. Instrum., 43, 555 (1972).

158. R. Freeman and H. D. W. Hill, J. Chem. Phys., 55, 1985 (1971).

159. T. K. Leipert, J. H. Noggle, W. J. Freeman, and D. L. Dalrymple, J. Magn. Reson., 19, 208 (1975).

160. T. K. Leipert, W. J. Freeman, and J. H. Noggle, J. Chem. Phys., 63, 4177 (1975).

161. W. J. Freeman, T. K. Leipert, D. L. Dalrymple, and J. H. Noggle, Rev. Sci. Instrum., 47, 146 (1976).

162. T. L. James, G. B. Matson, I. D. Kuntz, R. W. Fisher, and D. H. Buttlaire, J. Magn. Reson., 28, 417 (1977).

INDEX